N. A. Shaposhnikov and N. K. Valtsov

Collection of Algebra Problems

Part Two

1,531 Classical Algebra Problems

Translated by

Valery Manokhin, PhD

Northern Star Academic Press

Collection of Algebra Problems, Part Two

Published by Northern Star Academic Press
russianmathbooks.com

ISBN 978-1-919012-50-6 (paperback)
ISBN 978-1-919012-51-3 (hardcover)

Typeset in LaTeX
First English edition: 2026

Publisher's Note

This volume presents the second part of the algebraic problem collection by N. A. Shaposhnikov and N. K. Valtsov, prepared for advanced secondary-school study and independent practice.

Organization. Part Two is arranged in sixteen chapters, covering irrational expressions, equations, progressions, logarithms, combinatorics, probability, and related topics. The sequence is cumulative: earlier techniques are repeatedly reused in more demanding settings. Selected answers are provided at the end of the volume for checking work.

Asterisk. An asterisk marks a more difficult or optional problem.
Paired variants. A single base number may contain two variants. In this edition they are labelled *a* and *b*, so that the two exercises can never be confused. Answer entries retain the numbered structure of the exercises; solve each labelled variant separately when the source gives more than one form. Where a base number carries two variants, the 1935 key prints a single answer, and that answer belongs to the first variant, *a*. Variant *b* is therefore usually left unanswered in the original. Nothing has been supplied in its place: an answer the authors did not write would not be theirs, and would be impossible to distinguish from one that is.

About the authors. Nikolai Aleksandrovich Shaposhnikov (1851–1920) and Nikolai Konstantinovich Valtsov (1858–1900) belonged to a generation of Russian educators who treated a problem book as an instrument for developing mathematical thought, not merely as a store of exercises.

Shaposhnikov studied mathematics at the Fourth Moscow Gymnasium under Alexander Fyodorovich Malinin, whose books Northern Star Academic Press has also brought to English readers. After completing the gymnasium with a gold medal, he studied at Imperial Moscow University, earned a master's degree in pure mathematics, and later taught at the Imperial Moscow Technical School. His work joined serious mathematical scholarship to a lifelong concern for clear explanation and purposeful practice.

Valtsov was also trained at Imperial Moscow University. He taught mathematics and physics, first in Kolomna and later at the Fourth Moscow Gymnasium, and was remembered for the care he gave to pupils who needed additional help after lessons. His daily classroom experience complemented Shaposhnikov's work as a mathematician and established author.

Their algebra collection began appearing in 1887–1890. It had reached its twenty-fourth edition by 1917, eventually passed fifty-two editions, and remained in use for more than seventy years. It was also recognized with the Peter the Great Prize. That endurance reflects the quality still visible in this volume: careful sequencing, abundant variation, and sustained practice designed to turn understanding into fluency.

Contents

Publisher's Note **iii**

IX Irrational Expressions **1**

General Information on Roots. Extraction of a Root from a Monomial . . 1
Bringing a Factor Out from Under the Radical and Introducing a Factor Under the Radical . . . 3
Reducing Root Indices and Reducing Radicals to a Common Index 4
Reducing Roots to Normal Form . . . 5
Similarity of Roots . . . 6
Addition and Subtraction of Roots . . . 7
Multiplication and Division of Roots . . . 8
Raising Roots to a Power and Extracting a Root from Them . . . 11
Removing Irrationality in the Denominator of a Fraction . . . 12
Square Root of a Binomial of the Form $A \pm \sqrt{B}$ . . . 13
Problems on All Operations with Radicals . . . 14
Powers and Roots with Fractional Exponents . . . 15
Imaginary Numbers . . . 17

X Functions and Their Graphs **21**

XI Quadratic Equations **25**

Solving Literal Quadratic Equations . . . 25
Properties of the Roots of a Quadratic Equation . . . 26
Composing Literal Quadratic Equations . . . 29
Graph of a Quadratic Function . . . 31

XII Equations of Higher Degrees **35**

The Biquadratic Equation . . . 35
The Binomial Equation . . . 36
The Trinomial Equation . . . 36
Equations Whose Left-Hand Side Factors . . . 37
The Reciprocal, or Symmetric, Equation . . . 37

XIII Irrational Equations **41**

XIV Systems of Equations of Degree Higher than the First **43**
Solving Systems of Equations . . . 43
Composing Systems of Equations . . . 50
Graphical Solution of Systems of Two Equations of the Second Degree in Two Unknowns . . . 54

XV Progressions **55**
The Arithmetic, or Difference, Progression . . . 55
The Geometric, or Multiple, Progression . . . 60

XVI Logarithms **67**
General Properties of Logarithms . . . 67
Decimal Logarithms . . . 72
Exponential and Logarithmic Equations . . . 76
Compound Interest Problems . . . 79

XVII Combinatorics **81**

XVIII Newton's Binomial **85**

XIX Divisibility of Polynomials **87**

XX Inequalities **89**

XXI Solving Indeterminate Equations of the First Degree **93**

XXII Continued Fractions **97**

XXIII Investigation of Equations **99**
Investigation of First-Degree Equations with One Unknown . . . 99
Investigation of a System of First-Degree Equations with Two Unknowns 100
Investigation of Second-Degree Equations . . . 101

XXIV Limits **103**

Answers **105**
Chapter IX. Irrational Expressions . . . 105
Chapter X. Functions and Their Graphs . . . 117
Chapter XI. Quadratic Equations . . . 117
Chapter XII. Equations of Higher Degrees . . . 121
Chapter XIII. Irrational Equations . . . 122
Chapter XIV. Systems of Equations of Degree Higher than the First . . . 123
Chapter XV. Progressions . . . 127
Chapter XVI. Logarithms . . . 131
Chapter XVII. Combinatorics . . . 138
Chapter XVIII. Newton's Binomial . . . 139

Chapter XIX. Divisibility of Polynomials 139
Chapter XX. Inequalities 140
Chapter XXI. Indeterminate Equations of the First Degree 141
Chapter XXII. Continued Fractions 143
Chapter XXIII. Investigation of Equations 144
Chapter XXIV. Limits 145

Chapter IX

Irrational Expressions

General Information on Roots. Extraction of a Root from a Monomial

Definition. The root of the n-th degree of a is a quantity x which, raised to the n-th power, gives a. This quantity is expressed by the symbol $\sqrt[n]{a}$. In the equality $\sqrt[n]{a} = x$, a is called the radicand, n the index of the root, and x (or the equal $\sqrt[n]{a}$) the root of the n-th degree of a. The operation of finding x from given a and n is called extracting the root.

Rule of signs. An even-degree root of a positive quantity has two signs, positive and negative: $\sqrt[2n]{+a} = \pm \sqrt[2n]{a}$. An even-degree root of a negative quantity is an imaginary quantity. An odd-degree root of any quantity, positive or negative, has the same sign as the radicand: $\sqrt[2n+1]{+a} = + \sqrt[2n+1]{a}$, $\sqrt[2n+1]{-a} = - \sqrt[2n+1]{a}$.

Theorem 1. The root of a product equals the product of the roots of each factor: $\sqrt[n]{ab} = \sqrt[n]{a} \cdot \sqrt[n]{b}$. *Theorem 2.* The root of a fraction equals the root of the numerator divided by the root of the denominator: $\sqrt[n]{\frac{a}{b}} = \frac{\sqrt[n]{a}}{\sqrt[n]{b}}$. *Theorem 3.* The root of a power is obtained by dividing the exponent of the power by the index of the root: $\sqrt[n]{a^{mn}} = a^m$.

Here the roots of numerical coefficients are extracted directly, while the third theorem is applied to literal expressions. For example, we have

$$\sqrt[3]{\frac{27a^6b^3}{64c^{3n}d^{15}}} = \frac{3a^2b}{4c^nd^5}.$$

To extract a root from a monomial, one places the sign by the rule of signs, then extracts the required root from each factor and divisor, and arranges the results as factors or divisors correspondingly. The index of a root may be a negative quantity; every root with a negative index equals unity divided by the like root with positive index: $\sqrt[-n]{a} = \frac{1}{\sqrt[n]{a}}$. To roots with negative indices there apply those rules indicated above: the rule of signs, all three theorems, and the rule for extracting a root from a monomial.

Extract the root from the monomials:

1a. $\sqrt[6]{2^{12}}$.

1b. $\sqrt[4]{3^{8}}$.

2a. $\sqrt[3]{-a^{6}}$.

2b. $\sqrt{-10^{10}}$.

3a. $\sqrt[n]{a^{3n}}$.

3b. $\sqrt[3n]{a^{6n+9mn}}$.

4a. $\sqrt[n+2]{a^{3n+6}}$.

4b. $\sqrt[3+n]{a^{15+5n}}$.

5a. $\sqrt[3]{8 \cdot 3^{3}}$.

5b. $\sqrt[5]{32 \cdot 10^{5}}$.

6a. $\sqrt[4]{16 \cdot 81}$.

6b. $\sqrt[3]{125 \cdot 1000}$.

7a. $\sqrt{\frac{a^4}{9}}$.

7b. $\sqrt[3]{-\frac{a^3}{64}}$.

8a. $\sqrt[5]{-\frac{a^{10}}{b^{15}}}$.

8b. $\sqrt[7]{\frac{a^{21}}{b^{14}}}$.

9a. $\sqrt[4]{a^{16}b^{8}c^{4}}$.

9b. $\sqrt{2^{4}a^{6}b^{12}}$.

10a. $\sqrt[3]{-27a^{12}b^{3}}$.

10b. $\sqrt[5]{-32a^{5}b^{10}}$.

11a. $\sqrt[-3]{27}$.

11b. $\sqrt[-5]{32}$.

12a. $\sqrt[-2]{\frac{4}{9}}$.

12b. $\sqrt[-3]{\frac{27}{8}}$.

13a. $\sqrt[3]{a^{-6}}$.

13b. $\sqrt[3]{a^{-12}}$.

14a. $\sqrt[5]{-a^{-20}}$.

14b. $\sqrt[7]{-a^{-14}}$.

15a. $\sqrt[-5]{-\frac{1}{32}}$.

15b. $\sqrt[-3]{-\frac{1}{64}}$.

16a. $\sqrt[-n]{-\frac{1}{a^{n}}}$.

16b. $\sqrt[-n]{-\frac{1}{a^{3n}}}$.

17a. $\sqrt[4]{16a^{-4}b^{12}}$.

17b. $\sqrt[6]{64a^{-12}b^{6}}$.

18a. $\sqrt[-3]{\frac{8}{125}a^{3n}b^{-6}}$.

18b. $\sqrt[-4]{\frac{1}{81}a^{-8n}b^{4}}$.

19a. $\sqrt{6\frac{1}{4}a^{6}c^{4}m}$.

19b. $\sqrt{1\frac{11}{25}a^{4}b^{10n}}$.

20a. $\sqrt[4]{\frac{16}{81}a^{8n}b^{16}}$.

20b. $\sqrt[3]{\frac{125}{64}a^{6n}c^{15}}$.

21a. $\sqrt[3]{0.027a^{6n-3}b^{18}c^{-6}}$.

21b. $\sqrt[4]{0.0625a^{4n+8}b^{24}c^{-12}}$.

22a. $\sqrt[5]{-10^{10}a^{-20n}b^{5-15m}}$.

22b. $\sqrt[3]{-64a^{3n-6}b^{15n}}$.

23. $\sqrt{\frac{4^{-1}a^{4}b^{-6}}{9^{-1}c^{8}d^{-2}}}$.

24. $\sqrt[3]{\frac{343a^{-15}b^{18}}{2^{-6}c^{9}d^{-3}}}$.

25. $\sqrt[-2]{\frac{a^{2}b^{2n-6}c^{-2m}}{4a^{-6}f^{-4n+2}}}$.

26. $\sqrt[-3]{-\frac{1000p^{12}q^{-6}f^{3}t}{27a^{-3m}b^{9}}}$.

27. $\sqrt[9]{2^{36}a^{-40}b^{7}\frac{(a+b)^{7}}{a^{-4}b^{-11}}}$.

28. $2ab^{2}\sqrt{2a^{3}bc^{2}\sqrt[3]{8a^{3}b^{9}c^{6}}}$.

29. $\sqrt[-n]{\frac{(3a^{3y}-2)^{2n}a^{-(p+n)}b^{-(n+np)}c^{n}}{a-p}}$.

30. $3a^{5-n}b^{-4n}\sqrt[-3]{\frac{27}{64}a^{-15}b^{3n}c^{6-3n}d^{9}}$.

Bringing a Factor Out from Under the Radical and Introducing a Factor Under the Radical

If the radicand factors into two factors, one of which is a perfect power and the other an imperfect one, then one can extract the root from the first factor and multiply the obtained rational expression by the irrational root of the second. This transformation is called *bringing a factor out from under the radical.*

31a. $\sqrt{8}$.

31b. $\sqrt{18}$.

32a. $\sqrt{75}$.

32b. $\sqrt{28}$.

33a. $\sqrt[3]{81}$.

33b. $\sqrt[3]{500}$.

34a. $\sqrt[3]{-108}$.

34b. $\sqrt[3]{-72}$.

35a. $\sqrt[4]{48}$.

35b. $\sqrt[4]{162}$.

36a. $\sqrt[4]{1250}$.

36b. $\sqrt[4]{112}$.

37a. $\sqrt[5]{486}$.

37b. $\sqrt[5]{96}$.

38a. $\sqrt[5]{-224}$.

38b. $\sqrt[5]{-1215}$.

39a. $2\sqrt{405}$.

39b. $3\sqrt[3]{192}$.

40a. $\frac{2}{3}\sqrt[4]{243}$.

40b. $\frac{5}{2}\sqrt[5]{128}$.

41a. $\sqrt[4]{a^8c^3}$.

41b. $\sqrt[6]{a^{12}c^5}$.

42a. $\sqrt[5]{a^{15}b^6}$.

42b. $\sqrt[3]{a^9b^4}$.

43a. $\sqrt[3]{x^4y^5}$.

43b. $\sqrt[3]{x^{10}y^7}$.

44a. $\sqrt[4]{a^5b^6}$.

44b. $\sqrt[4]{a^{10}b^7}$.

45a. $\sqrt{4a^4b}$.

45b. $\sqrt{25ab}$.

46a. $\sqrt[3]{64x^6y^4}$.

46b. $\sqrt[3]{27x^8y^3}$.

47a. $3\sqrt{80c^4d^2}$.

47b. $2\sqrt{75c^6d^4}$.

48a. $2\sqrt{\frac{a^5}{4}}$.

48b. $3\sqrt{\frac{a^4}{27}}$.

49a. $\sqrt[3]{\frac{a^8}{b^9}}$.

49b. $\sqrt[5]{\frac{a^{14}}{b^{10}}}$.

50a. $\sqrt[6]{\frac{a^5}{b^{18}}}$.

50b. $\sqrt[4]{\frac{a^7}{b^{16}}}$.

51a. $a\sqrt{\frac{0.54z}{a^2x^2}}$.

51b. $a^2\sqrt{-\frac{0.54z}{a^6x^6}}$.

52a. $\sqrt[3]{\frac{-0.729m}{a^6}}$.

52b. $\sqrt{\frac{8.64m}{a^4}}$.

53a. $\sqrt{\frac{(a^2-2ab+b^2)y}{25}}$.

53b. $\sqrt{\frac{50z}{a^2+2ab+b^2}}$.

54a. $\sqrt{\frac{a}{b^2}-\frac{1}{b}}$.

54b. $\sqrt[3]{\frac{x^3}{a^6}-\frac{1}{a}}$.

55a. $\sqrt[3]{\frac{(y^2-x^2)^4}{8(x+y)}}$.

55b. $\sqrt[5]{\frac{(x^2-y^2)^6}{32(y-x)}}$.

56a. $a\sqrt[3]{\frac{b^3}{a^4}-\frac{b^5}{a^6}}$.

56b. $\frac{3}{2a}\sqrt{4a^2-\frac{8a^2b^2}{9}}$.

57a. $\sqrt[m]{2^{m+1}a^{5m}b^{m+n}c^{mp+1}}$.

57b. $\sqrt[m+n]{a^{2m+n}b^{m+2n}c^{m^2-n^2}}$.

58. $yz^2\sqrt[2r]{x^{4r+1}y^{6r+2}z^5}$.

59. $\frac{ac}{b}\cdot\sqrt[n]{3^{n+2}a^{n+5}b^{n-1}c^{1-3n}}$.

If a rational factor stands before the root, one can introduce it under the radical by raising it to the power indicated by the index of the root and multiplying the result by the radicand. This is called *introducing a factor under the radical.*

60a. $2\sqrt{3}$. **60b.** $3\sqrt{2}$.

61a. $6\sqrt{5}$. **61b.** $4\sqrt{3}$.

62a. $3\sqrt[3]{2}$. **62b.** $2\sqrt[3]{3}$.

63a. $5\sqrt[3]{3}$. **63b.** $7\sqrt[3]{2}$.

64a. $2\sqrt[5]{5}$. **64b.** $3\sqrt[5]{4}$.

65a. $a\sqrt{5}$. **65b.** $5\sqrt{a}$.

66a. $x\sqrt[4]{2}$. **66b.** $y\sqrt[6]{5}$.

67a. $5\sqrt[4]{a}$. **67b.** $a\sqrt[4]{5}$.

68a. $m\sqrt[3]{n}$. **68b.** $n\sqrt[5]{m^2}$.

69a. $n^2\sqrt{a}$. **69b.** $m\sqrt{a}$.

70a. $3a\sqrt{ax}$. **70b.** $a^3\sqrt{2ab}$.

71a. $m^2\sqrt[3]{mn}$. **71b.** $2n\sqrt[3]{m^2n}$.

72a. $\frac{1}{2}\sqrt{a}$. **72b.** $\frac{2}{3}\sqrt[3]{a^2}$.

73a. $\frac{x}{y}\sqrt[3]{y^2}$. **73b.** $\frac{y}{x}\sqrt[3]{\frac{x}{y}}$.

74a. $\frac{a}{b}\sqrt[3]{-\frac{b^4}{a^5}}$. **74b.** $-\frac{b}{a}\sqrt[5]{-\frac{a^2}{b^3}}$.

75a. $m\sqrt[5]{1-\frac{1}{m^5}}$. **75b.** $\frac{1}{m}\sqrt[4]{m^5-1}$.

76a. $(m+n)\sqrt{\frac{1}{m^2-n^2}}$. **76b.** $\frac{1}{m-n}\sqrt{m^2-n^2}$.

77a. $2ac^3\sqrt[3]{3abc^2}$. **77b.** $\frac{4a}{3b}\sqrt[5]{\frac{27b^3}{16a^4}}$.

78a. $3a^nb\cdot\sqrt[m]{3a^2b}$. **78b.** $2ab^m\cdot\sqrt[n]{3a^mb^2}$.

79a. $3a^2c^3\sqrt[4]{2a^nb^{-3}}$. **79b.** $2a^nb^{-2}\sqrt[3]{5a^{-n}b^3}$.

Reducing Root Indices and Reducing Radicals to a Common Index

The value of a root is unchanged if one multiplies or divides the index of the root and the exponent of the radicand by one and the same number. Two corollaries follow: 1) if the index of the root and the exponent of the radicand have a common factor, they can be reduced by it; 2) if several roots have different indices, then by multiplying the indices and the exponents of the radicands respectively by suitable numbers, the roots can be reduced to a common index. To multiply the exponent of the radicand means the same as raising that expression to the power corresponding to the multiplier; to divide the exponent of the radicand means the same as extracting from that expression the root corresponding to the divisor.

Reduce the root indices:

80a. $\sqrt[9]{a^6}$.

80b. $\sqrt[6]{a^4}$.

81a. $\sqrt[8]{a^{10}b^{12}}$.

81b. $\sqrt[10]{a^{15}b^{25}}$.

82a. $\sqrt[3n]{a^{2n}b^{3n}}$.

82b. $\sqrt[5n]{a^{10}b^{5n}}$.

83a. $\sqrt[nm]{a^m b^{2m}}$.

83b. $\sqrt[mn]{a^{2mn}b^{3n}}$.

84a. $\sqrt[6]{9a^4b^6}$.

84b. $\sqrt[4]{4a^8b^2}$.

85a. $\sqrt[9]{27a^{3m}b^6}$.

85b. $\sqrt[12]{64a^9b^{3m}}$.

86a. $\sqrt[12]{64a^4b^{2n}}$.

86b. $\sqrt[18]{81a^{16}b^{4n}}$.

87a. $\sqrt[6n]{\frac{16a^{10}b^{-6}}{9c^{18}}}$.

87b. $\sqrt[6n]{\frac{27a^{-9}b^{12}}{8c^{15}}}$.

88a. $\sqrt[12]{\frac{1000a^{-6}}{729b^9c^{-3}}}$.

88b. $\sqrt[8]{\frac{16a^4b^{12}}{81c^{-6}}}$.

89a. $\sqrt[-4]{a^{-8}b^{10}c^{-2}}$.

89b. $\sqrt[-6]{9a^4b^{-8}c^4}$.

Reduce the roots to a common index:

90a. $\sqrt[6]{a^5}$ and $\sqrt[4]{a^3}$.

90b. $\sqrt[9]{a^4}$ and $\sqrt[6]{a^5}$.

91a. $\sqrt[3]{2a^2}$ and $\sqrt[6]{ab^5}$.

91b. $\sqrt{3a}$ and $\sqrt[4]{2b^3}$.

92a. $\sqrt[3]{2a^2b}$ and $\sqrt[4]{3a^3b}$.

92b. $\sqrt[5]{3a^3b^2}$ and $\sqrt[3]{2ab}$.

93a. $\sqrt[12]{\frac{3a^5}{b^3}}$ and $\sqrt[9]{\frac{10b^2}{a}}$.

93b. $\sqrt[8]{\frac{5a}{b^2}}$ and $\sqrt[6]{\frac{3a^2}{b}}$.

94a. $\sqrt[m^2]{\frac{3a^2}{bc^3}}$ and $\sqrt[mn]{\frac{2ab^2}{c^3}}$.

94b. $\sqrt[mn]{\frac{2b^3}{ac^4}}$ and $\sqrt[n^2]{\frac{3a^2c}{b^2}}$.

95a. $\sqrt[12]{a^2b^3}$, $\sqrt[4]{a}$ and $\sqrt[8]{a^3}$.

95b. $\sqrt[6]{ab^4}$, $\sqrt[9]{a^4}$ and $\sqrt[12]{b^5}$.

96a. $\sqrt[6]{a^2b}$, $\sqrt[15]{a^3b^4}$ and $\sqrt[50]{a^{10}b^{20}}$.

96b. $\sqrt[8]{a^4b^5}$, $\sqrt[12]{a^7b^3}$ and $\sqrt[15]{a^{10}b^{25}}$.

97a. $\sqrt{\frac{x}{y}}$, $\sqrt[5]{\frac{y^3}{z^2}}$ and $\sqrt[3]{\frac{a^2}{b}}$.

97b. $\sqrt{\frac{a^3}{b^2}}$, $\sqrt[5]{\frac{x}{y^4}}$ and $\sqrt[8]{\frac{y}{z^4}}$.

98. $\sqrt[4]{\frac{x-1}{x+1}}$, $\sqrt[2n]{\frac{x+1}{x-1}}$ and $\sqrt[n]{\frac{x}{y}}$.

99. $\sqrt[2n]{\frac{a+b}{x}}$, $\sqrt[6]{\frac{a}{x+y}}$ and $\sqrt[3n]{\frac{a}{b}}$.

Reducing Roots to Normal Form

Every root can be reduced to its simplest, or normal, form. For this one carries out successively: transform the radicand into a monomial if possible; reduce the index of the root if it has a common factor with the exponents of all factors and divisors of the radicand; bring out from under the radical that part of the radicand which admits root extraction; remove the irrationality of the denominator (multiplying numerator and denominator of the radicand by an expression chosen so that the denominator becomes a perfect power, then extracting the root from the denominator).

Reduce the following roots to simplest form:

100a. $\frac{3xy^2}{2}\sqrt[3]{\frac{8}{xy}}$.

100b. $\frac{2x}{3y^2}\sqrt{\frac{8y^3}{x^5}}$.

101. $a^2\sqrt{\frac{2ab^3}{3c^2d}}$.

102. $\frac{2ab}{c}\sqrt[3]{\frac{5a}{16b^2c^3}}$.

103. $\frac{1}{a}\sqrt[3]{a^8-a^6b^2}$.

104. $a^2\sqrt[4]{\frac{1}{a^3}-\frac{b}{a^4}}$.

105. $5n^x\sqrt[3]{\frac{ab^5}{25n^{3x+1}}}$.

106. $\frac{c^{n-m}}{a^m}\sqrt[m+n]{\frac{a^{m^2-n^2}\cdot b^{m+6n}}{c^{m+2n}}}$.

107. $\frac{a+b}{a}\sqrt[3]{\frac{a^{13}-a^{12}b}{(a-b)^2}}$.

108. $\frac{a}{c}\sqrt{\frac{a^3b-4a^2b^2+4ab^3}{c^2}}$.

Similarity of Roots

When a root is reduced to simplest form, the rational factor of the root is called its coefficient. Roots are called *similar* if they differ only in coefficients but have the same indices and the same radicands. To judge whether given roots are similar, one must reduce them to simplest form.

Prove the similarity of the roots:

109. $\sqrt{3}$ and $\sqrt{12}$.

110. $\sqrt{63}$ and $\sqrt{28}$.

111. $\sqrt[3]{54}$ and $2\sqrt[3]{2}$.

112. $\sqrt[4]{80}$ and $\sqrt[4]{405}$.

113. $\sqrt{18}$, $\sqrt{128}$ and $\sqrt{32}$.

114. $\sqrt[3]{54}$, $\sqrt[3]{16}$ and $\sqrt[3]{432}$.

115. $\sqrt{\frac{4}{3}}$ and $\sqrt{12}$.

116. $\sqrt{\frac{2}{5}}$ and $\sqrt{\frac{2}{45}}$.

117. $\frac{1}{4}\sqrt{0.2}$ and $\frac{1}{5}\sqrt{5}$.

118. $\sqrt[3]{\frac{8}{3}}$ and $\sqrt[3]{\frac{9}{8}}$.

119. $\sqrt{\frac{1}{2}+\frac{3}{4}}$ and $\sqrt{\frac{8}{9}-\frac{1}{3}}$.

120. $\sqrt[3]{\frac{6}{25}-\frac{1}{4}}$ and $\sqrt[3]{\frac{1}{27}-\frac{1}{32}}$.

121. $\sqrt[6]{a^7b}$ and $\sqrt[6]{a^{13}b^7}$.

122. $\sqrt[3]{0.027xy^2}$ and $\sqrt[3]{0.064\frac{x}{y}}$.

123. $\sqrt{a-\frac{1}{a^4}}$ and $\sqrt{\frac{a^3-1}{a^4}}$.

124. $\sqrt{\frac{1}{b}-a}$ and $\sqrt{\frac{bd^2-ab^2d^2}{c^2}}$.

125. $\sqrt{\left(\frac{a^2-b^2}{a+b}\right)^3}$, $\sqrt{\frac{(a^2+b^2)^2}{a-b}}$ and $\sqrt{a^3-a^2b}$.

126. $\frac{1}{y}\sqrt{x^2y\left(\frac{x}{y}-1\right)}$, $x\sqrt{\frac{z}{xz-yz}}$ and $\sqrt{\frac{4x}{y^2}-\frac{4}{y}}$.

127. $\sqrt[3]{8a^5-16a^3b^2}$, $ab\sqrt[3]{\frac{1}{a}-\frac{2b^2}{a^3}}$ and $\sqrt[3]{\frac{2}{a^3b}-\frac{1}{ab^3}}$.

128. $\frac{x^2}{y}\sqrt[n]{x^{-3(n-1)}y^{2n+1}}$, $\frac{1}{xy}\sqrt[n]{x^{n+3}y^{n+1}}$ and $(2x-y)\sqrt[n]{x^{3-n}y}$.

Addition and Subtraction of Roots

To add and subtract roots, one connects them by the signs of these operations, reduces the roots to normal form, and, if similar roots appear, performs the reduction: the coefficients of similar terms, taken with the signs of the corresponding terms, are enclosed in brackets and the common root is brought outside as a factor; the obtained common coefficient is then simplified.

129a.
$(5\sqrt{2}-4\sqrt[3]{3})+(3\sqrt{2}+6\sqrt[3]{3})$.

129b.
$(7\sqrt[3]{4}-2\sqrt{5})-(5\sqrt[3]{4}-4\sqrt{5})$.

130a.
$(10\sqrt[4]{7}+\sqrt[5]{3})-(5\sqrt[5]{3}+2\sqrt[4]{7})$.

130b.
$(2\sqrt[3]{11}-8\sqrt[5]{7})+(7\sqrt[5]{7}-\sqrt[3]{11})$.

131a.
$(a\sqrt{b}-b\sqrt{c})-(3a\sqrt{b}-5b\sqrt{c})$.

131b.
$(3\sqrt[3]{a}+b\sqrt{c})-(2\sqrt[3]{a}+3b\sqrt{c})$.

132a.
$(a\sqrt[5]{b^4}-2c\sqrt[4]{d})-(-5c\sqrt[4]{d}+3a\sqrt[5]{b^4})$.

132b.
$(2\sqrt[4]{a^3}-\sqrt[3]{a^2b})+(-\sqrt[4]{a^3}+5\sqrt[3]{a^2b})$.

133a.
$\sqrt{2}+3\sqrt{32}+\frac{1}{2}\sqrt{128}-6\sqrt{18}$.

133b.
$\sqrt{75}-\sqrt{147}+\sqrt{48}-\frac{1}{5}\sqrt{300}$.

134a.
$20\sqrt{245}-\sqrt{5}+\sqrt{125}-2\frac{1}{2}\sqrt{180}$.

134b.
$\sqrt{275}-10\sqrt{11}-2\sqrt{99}+\sqrt{396}$.

135a.
$\frac{1}{2}\sqrt[3]{5}-2\frac{1}{4}\sqrt[3]{40}+10\sqrt[3]{135}-\sqrt[3]{320}$.

135b.
$3\sqrt[5]{2}-\frac{1}{2}\sqrt[5]{64}+10\sqrt[5]{486}-6\frac{1}{2}\sqrt[5]{2}$.

136a.
$\sqrt{\frac{45}{4}}-\sqrt{20}-5\sqrt{\frac{1}{18}}-\frac{1}{6}\sqrt{245}-\sqrt{\frac{49}{2}}$.

136b.
$2\sqrt{\frac{5}{3}}-\sqrt{60}-\sqrt{15}+\sqrt{\frac{3}{5}}+\sqrt{\frac{4}{15}}$.

137a.
$3\frac{1}{2}\sqrt{24}-\frac{\sqrt[3]{54}}{4}+\frac{2\sqrt{99}}{3}-1\frac{1}{2}\sqrt{44}+3\sqrt[3]{2}$.

137b.
$\sqrt[3]{54}+\sqrt{\frac{1}{2}}-\sqrt[3]{250}-\frac{3}{4}\sqrt{\frac{2}{9}}+\sqrt{6\frac{3}{4}}$.

138a.
$5\sqrt{8}-8\sqrt{\frac{1}{3}}-\sqrt{4\frac{1}{2}}+6\sqrt{\frac{5}{3}-\frac{13}{9}}+\frac{4\sqrt{3}}{3}$.

138b.
$3\sqrt[3]{32}+\sqrt[3]{\frac{1}{9}}-\sqrt[3]{108}-16\sqrt[3]{\frac{1}{16}}+4\sqrt[3]{\frac{1}{72}}$.

139a. $\sqrt{a^3}+b\sqrt{a}-\sqrt{9a}$.

139b. $\sqrt[3]{a^2}+\sqrt[3]{\frac{a^5}{8}}-3a\sqrt[3]{\frac{1}{a}}$.

140a. $\sqrt[3]{27a^4}-3\sqrt[3]{8a}+3\sqrt[3]{125a^7}$.

140b. $\sqrt[5]{a^5b}-\sqrt[5]{32b^6}+3a\sqrt[5]{b}$.

141a.
$3\sqrt{125a^3b^2}+b\sqrt{20a^3}-\sqrt{500a^3b^2}$.

141b.
$2\sqrt[3]{a^6b}-3a^2\sqrt[3]{64b}+2a^2\sqrt[3]{125b^4}$.

142a. $\frac{1}{a^2c}\sqrt{3a^8c^4d}-\frac{2}{ac^2}\sqrt{12a^6c^6d}-a^4c^2\sqrt{\frac{3d}{a^4c^2}}$.

142b. $4ac^2\sqrt[3]{a^5b^7}+b^3\sqrt[3]{a^2b^4c^6}-\frac{3}{2}\sqrt[3]{8a^2b^{13}c^6}$.

143a. $5\sqrt[3]{x^2y^5} + 4y^2\sqrt[3]{\frac{x^2}{y}} + \frac{4y}{x^2}\sqrt[3]{-x^8y^2} - 6xy\sqrt[3]{\frac{y^2}{x}} - \frac{3}{2}xy^2\sqrt[3]{-\frac{8}{xy}}$.

143b. $\sqrt[3]{xy} + 6xy\sqrt[3]{\frac{1}{x^4y^2}} - 4x^2y^2\sqrt[3]{-\frac{1}{x^5y^5}} + \frac{1}{2}y\sqrt[3]{\frac{x}{y^4}} - \frac{3}{2x}\sqrt[3]{-x^4y}$.

144. $\sqrt{m^3 - m^2n} - \sqrt{(m+n)(m^2-n^2)} - \sqrt{mn^2 - n^3}$.

145. $\sqrt{1 - \frac{x}{2}} - 3\sqrt{4-2x} - \sqrt{16-8x} + 8\sqrt{\frac{1}{4} - \frac{x}{8}}$.

146. $(a^4 - 2b^4)\sqrt{\frac{a+b}{a-b}} - (a^2+b^2)\sqrt{(a+b)^3(a-b)} + \frac{b^2}{a-b}\sqrt{a^2b^4 - b^6}$.

147. $\frac{x}{2}\sqrt[4]{(1+2x+x^2)(x+1)(x^2-1)} - \sqrt[4]{x^5(1-x^{-1})} + \frac{1}{2}x^3\sqrt[4]{x^{-3} - x^{-4}}$.

148. $\sqrt[3]{8x^9 - 8x^6y^3} + x\sqrt[3]{x^3y^3 - x^6} + \sqrt[3]{1 - x^3y^{-3}} + \frac{x^2}{y^2}\sqrt[3]{x^{-3}y^3 - x^{-6}y^6}$.

Multiplication and Division of Roots

The product of roots with the same index equals the root of the same degree of the product of the radicands. The quotient of like roots equals the root of the same degree of the quotient of the radicands. If the indices differ, they are first reduced to a common index. Coefficients are multiplied or divided separately and the result written before the common root.

Carry out the indicated operations on the roots:

149a. $\sqrt{3}\cdot\sqrt{27}$.

149b. $\sqrt{5}\cdot\sqrt{20}$.

150a. $\sqrt[3]{2}\cdot\sqrt[3]{16}$.

150b. $\sqrt[3]{3}\cdot\sqrt[8]{18}$.

151a. $3\sqrt[3]{18}\cdot\frac{5}{6}\sqrt[3]{-6}$.

151b. $2\sqrt[3]{16}\cdot\frac{3}{4}\sqrt[3]{-5}$.

152a. $\frac{1}{3}\sqrt[4]{27}\cdot\frac{1}{9}\sqrt[4]{243}$.

152b. $\frac{1}{2}\sqrt[6]{32}\cdot\frac{1}{4}\sqrt[6]{128}$.

153a. $\sqrt[3]{-108}\cdot\sqrt[3]{50}\cdot\sqrt[3]{40}$.

153b. $\sqrt[5]{7^3}\cdot\sqrt[5]{-112}\cdot\sqrt[5]{14}$.

154a. $2\sqrt[4]{32}\cdot\sqrt[4]{216}\cdot 3\sqrt[4]{60}$.

154b. $\sqrt[6]{1024}\cdot 2\sqrt[6]{6561}\cdot\sqrt[6]{1620}$.

155a.
$\left(4\sqrt{8} + \frac{1}{12}\sqrt{12} - \frac{1}{4}\sqrt{32}\right)\cdot 8\sqrt{32}$.

155b.
$\left(2\sqrt[3]{135} - 5\sqrt[3]{5} - 10\sqrt[3]{15}\right)\cdot\frac{1}{2}\sqrt[3]{75}$.

156a.
$(\sqrt[3]{9} - 7\sqrt[3]{72} + 6\sqrt[3]{1125})\cdot 4\sqrt{\frac{1}{9}}$.

156b.
$\left(2\sqrt{\frac{2}{3}} - 8\sqrt{\frac{3}{8}} + 3\sqrt{\frac{3}{2}}\right)\cdot 3\sqrt{\frac{2}{3}}$.

157a.
$\left(3\sqrt{\frac{5}{6}} - 5\sqrt{30} - 2\sqrt{\frac{15}{2}}\right)\cdot 2\sqrt{\frac{3}{2}}$.

157b.
$\left(6\sqrt[3]{\frac{9}{4}} - 5\sqrt[3]{36} + 9\sqrt[3]{\frac{16}{81}}\right)\cdot\frac{4}{3}\sqrt[3]{\frac{4}{9}}$.

158a. $(2\sqrt{6} - 3\sqrt{5})\cdot(\sqrt{3} + 2\sqrt{2})$.

158b. $(\sqrt[3]{9} - 2\sqrt[3]{4})\cdot(4\sqrt[3]{3} + \sqrt[3]{2})$.

159a.
$(\sqrt[3]{16} - 2\sqrt[3]{2} + 4\sqrt[3]{54})\cdot\left(5\sqrt[3]{4} - 3\sqrt[3]{\frac{1}{2}}\right)$.

159b. $\left(\frac{2}{5}\sqrt[3]{25} + \frac{1}{5}\sqrt[3]{200} - \frac{1}{2}\sqrt[3]{75}\right)\cdot(2\sqrt[3]{5} - 5\sqrt[3]{15})$.

160a. $\left(3\sqrt{\frac{2}{3}} - \sqrt{12} - \sqrt{6}\right)\cdot\left(2\sqrt{\frac{2}{3}} - 8\sqrt{\frac{3}{8}} + 3\sqrt{\frac{3}{2}}\right)$.

160b. $\left(5\sqrt[3]{\frac{4}{3}} - 3\sqrt[3]{\frac{3}{8}} + 4\sqrt[3]{\frac{2}{9}}\right)\cdot\left(6\sqrt[3]{\frac{9}{4}} - \sqrt[3]{36} - \sqrt[3]{72}\right)$.

161a. $\sqrt{a^3b}\cdot\sqrt{a^5b^2}$.

161b. $\sqrt[3]{a^2b}\cdot\sqrt[3]{ab^4}$.

162a. $a^2\sqrt[3]{2x}\cdot\frac{1}{a}\sqrt[3]{4x}$.

162b. $\frac{1}{a}\sqrt[4]{4x^2}\cdot a^3\sqrt[4]{8x}$.

163a. $2\sqrt[3]{25a^5}\cdot 3\sqrt[3]{15a^4}$.

163b. $5\sqrt[3]{12a^2}\cdot 2\sqrt[3]{18a^5}$.

164a. $3\sqrt{\frac{5a}{b^2}} \cdot 2\sqrt{\frac{4b^4}{5a^3}}$.

164b. $\frac{1}{2}\sqrt{\frac{8a}{3b^2}} \cdot \frac{3}{4}\sqrt{\frac{3b^3}{2a^3}}$.

165a. $\frac{x}{a}\sqrt[3]{\frac{a^2}{x}} \cdot \frac{1}{4}a\sqrt[3]{\frac{8a}{x^4}}$.

165b. $5\sqrt[3]{\frac{2.4}{25x^3}} \cdot \sqrt[8]{\frac{4a^5}{5x^2}}$.

166a. $\frac{12a^3}{5x^2}\sqrt[4]{\frac{a^7x}{32}} \cdot \frac{10x^3}{3a^2}\sqrt[4]{\frac{4}{a^3x}}$.

166b. $\frac{x^2}{a^2}\sqrt[3]{\frac{3a}{x^2}} \cdot \frac{1}{a^2x^5}\sqrt[3]{\frac{x^3}{a^4}}$.

167a.
$\sqrt[3]{\frac{3a^{-2}b^5}{5a^4b^{-2}}} \cdot \sqrt[3]{\left(\frac{6a^{-2}}{5b^3}\right)^{-2}} \cdot \sqrt[3]{-60a^5b^2}$.

167b. $\sqrt[3]{\left(\frac{2a^{-3}b}{9a^3b^{-1}}\right)^{-2}} \cdot$
$\sqrt[3]{\left(-\frac{3b^{-4}}{4a^{-5}}\right)^{-1}} \cdot \sqrt[3]{72a^4b^6}$.

168a. $\left(\sqrt{a} + \sqrt{ab} - \sqrt{\frac{a}{b}}\right) \cdot \sqrt{\frac{a}{b}}$.

168b.
$\left(\sqrt[3]{a^2b} + \sqrt[3]{\frac{b^2}{a^2}} - \sqrt[3]{ab^2}\right)\sqrt[3]{\frac{a^2}{b^2}}$.

169a. $\left(\sqrt{a} + \sqrt{\frac{b}{a}}\right) \cdot \left(\sqrt{ab} - \sqrt{\frac{a}{b}}\right)$.

169b. $\left(a + \frac{2}{a}\sqrt{ab}\right) \cdot \left(a - 2\sqrt{\frac{b}{a}}\right)$.

170a.
$\left(\sqrt[3]{a^2b} + \sqrt[3]{ab^2}\right) \cdot \left(\sqrt[3]{a} - \sqrt[3]{b}\right)$.

170b.
$\left(\sqrt[3]{a^2b^2} - \sqrt[3]{ab}\right) \cdot \left(\sqrt[3]{\frac{a}{b^2}} - \sqrt[3]{\frac{b}{a^2}}\right)$.

171a. $\sqrt{3} \cdot \sqrt[3]{2}$.

171b. $\sqrt[3]{5} \cdot \sqrt{2}$.

172a. $\sqrt[5]{\frac{3}{8}} \cdot \sqrt[3]{\frac{2}{3}}$.

172b. $\sqrt[3]{\frac{2}{5}} \cdot \sqrt[4]{\frac{3}{4}}$.

173a. $\sqrt[6]{54} \cdot \sqrt{6} \cdot \sqrt[3]{2}$.

173b. $\sqrt[3]{2} \cdot \sqrt[6]{\frac{1}{3}} \cdot \sqrt[4]{\frac{3}{4}}$.

174a. $\sqrt[9]{\frac{9}{4}} \cdot \sqrt[4]{\frac{2}{3}} \cdot \sqrt[6]{2} \cdot \sqrt[12]{3}$.

174b. $\sqrt[7]{12} \cdot \sqrt[3]{\frac{1}{2}} \cdot \sqrt[14]{6} \cdot \sqrt[6]{3}$.

175a.
$\left(3\sqrt[7]{10} - 2\sqrt[3]{4} + 6\sqrt[6]{25}\right) \cdot \sqrt[4]{2}$.

175b.
$\left(2\sqrt{6} + 3\sqrt[3]{15} - \sqrt[5]{10}\right) \cdot \sqrt[4]{12}$.

176a.
$\left(2\sqrt[7]{10} + 3\sqrt{2} - 4\sqrt[3]{5}\right) \cdot \sqrt[4]{10}$.

176b. $\left(2\sqrt{2} + 3\sqrt[3]{2} - 4\sqrt[4]{2}\right) \cdot 3\sqrt[6]{2}$.

177a.
$\left(3\sqrt{2} + 4\sqrt[3]{3}\right) \cdot \left(\sqrt{2} - 2\sqrt[3]{3}\right)$.

177b. $\left(5\sqrt[4]{3} - 2\sqrt[3]{2}\right) \cdot \left(\sqrt[3]{2} - \sqrt{3}\right)$.

178a.
$\left(6\sqrt[3]{2} - \sqrt[6]{32}\right) \cdot \left(\frac{3}{2}\sqrt[3]{2} - 2\sqrt[6]{\frac{1}{2}}\right)$.

178b.
$\left(\sqrt[3]{4} - 2\sqrt[4]{8}\right) \cdot \left(2\sqrt[3]{2} - 3\sqrt[4]{\frac{1}{2}}\right)$.

179a. $\sqrt[4]{a^3b} \cdot \sqrt[6]{ab^4}$.

179b. $\sqrt[12]{a^3b^5} \cdot \sqrt[9]{ab^2}$.

180a. $3a^2b\sqrt{3bc} \cdot 5ab\sqrt[3]{2a^2c}$.

180b. $8a^2b\sqrt[3]{3ac^4} \cdot 2ac^2\sqrt[4]{2b^2c}$.

181a.
$a^2\sqrt[4]{a^5b^2} \cdot b\sqrt[3]{\frac{a^5}{b}} \cdot \sqrt[4]{a^6b^7} \cdot ab\sqrt[3]{a^4b^7}$.

181b.
$a\sqrt[5]{a^4b^3} \cdot ab^2\sqrt[3]{ab^2} \cdot \sqrt[5]{ab^4} \cdot a\sqrt[3]{\frac{b^4}{a}}$.

182. $\left(\sqrt[3]{a^2} - 2\sqrt[4]{b^2} - a\sqrt[6]{b^5}\right) \cdot a^2\sqrt{ab}$.

183. $(\sqrt[5]{a^2} - \sqrt[3]{a^4} + a\sqrt{a^3}) \cdot (-2a\sqrt[3]{a^2})$.

184. $(\sqrt{a} - \sqrt[3]{a^2} + \sqrt[5]{a^4}) \cdot (\sqrt[3]{a} - \sqrt[15]{a^4})$.

185a. $\sqrt{28} : \sqrt{7}$.

185b. $\sqrt{45} : \sqrt{5}$.

186a. $\frac{\sqrt[3]{81}}{\sqrt[3]{3}}$.

186b. $\frac{\sqrt[3]{256}}{\sqrt[3]{4}}$.

187a. $\sqrt{\frac{12}{35}} : \sqrt{\frac{7}{5}}$.

187b. $\sqrt{\frac{10}{3}} : \sqrt{\frac{3}{5}}$.

188a. $\frac{3}{2}\sqrt[3]{96} : 3\sqrt[3]{\frac{3}{4}}.$

188b. $2\sqrt[3]{\frac{4}{25}} : \frac{1}{3}\sqrt[3]{\frac{2}{125}}.$

189a. $(5\sqrt[3]{4} - 6\sqrt[3]{10} + 15\sqrt[3]{16}) : 3\sqrt[3]{\frac{1}{2}}.$

189b. $(3\sqrt[3]{6} + 2\sqrt[3]{18} - 4\sqrt[3]{12}) : 2\sqrt[3]{\frac{2}{3}}.$

190a. $\sqrt{5a} : \sqrt{a}.$

190b. $\sqrt[3]{3a^2} : \sqrt[3]{a}.$

191a. $\sqrt[3]{4a^8} : \sqrt[3]{2a^2}.$

191b. $\sqrt{2a^3} : \sqrt{2a}.$

192a. $\sqrt[4]{27a^3} : \sqrt[4]{\frac{a^2}{3}}.$

192b. $\sqrt[4]{\frac{3a^3}{2}} : \sqrt[4]{\frac{8}{27a^3}}.$

193a. $\sqrt[4]{\frac{8a^5}{3b}} : \sqrt[4]{\frac{6x}{b^3}}.$

193b. $\sqrt[4]{\frac{3}{a^3}} : \sqrt[4]{\frac{4b^2}{3a^5}}.$

194a. $(ab^2\sqrt{x} - x\sqrt{b}) : \sqrt{bx}.$

194b. $(2 \cdot b\sqrt[3]{x^2} - x\sqrt[3]{b}) : \sqrt[3]{bx}.$

195. $\sqrt[4]{a^3x^3} - x\sqrt[4]{a^3} - 4a\sqrt[4]{ax^2}) : \sqrt[4]{ax^3}.$

196. $\left(2\sqrt[4]{x^3y} - 3\sqrt[4]{\frac{xy^3}{2}} + \sqrt[4]{\frac{1}{x}}\right) : \frac{1}{xy}\sqrt[4]{x^3y^2}.$

197. $(\sqrt[3]{a^2} - \sqrt[3]{b^2}) : (\sqrt[3]{a} + \sqrt[3]{b}).$

198. $(\sqrt[3]{a^2b} - 2\sqrt[3]{2ab^2} + b\sqrt[3]{4}) : (\sqrt[3]{a} - \sqrt[3]{2b}).$

199. $(\sqrt[4]{8a^3} - b\sqrt[4]{27b^2}) : (\sqrt[4]{2a} - \sqrt[4]{3b^2}).$

200. $(a^2\sqrt[4]{a} + b^2\sqrt[4]{8b}) : (\sqrt[4]{a^3} + \sqrt[4]{2b^3}).$

201. $(x^2\sqrt[3]{x^2} + xy\sqrt[3]{xy} + y^2\sqrt[3]{y^2}) : (x\sqrt[3]{x} + \sqrt[3]{x^2y^2} + y\sqrt[3]{y}).$

202a. $\sqrt[3]{9} : \sqrt{3}.$

202b. $\sqrt[5]{8} : \sqrt[3]{2}.$

203a. $\sqrt[5]{\frac{4}{5}} : 2\sqrt{\frac{1}{400}}.$

203b. $\sqrt{\frac{3}{5}} : \frac{1}{3}\sqrt[6]{3\frac{3}{8}}.$

204a. $(\sqrt[4]{6} - 2\sqrt{3} + \sqrt[3]{6}) : \frac{1}{2}\sqrt{6}.$

204b. $(3\sqrt{2} - 12\sqrt[3]{12} + 10\sqrt[4]{2}) : \frac{2}{3}\sqrt[4]{2}.$

205a. $\sqrt{a} : \sqrt[3]{a^2}.$

205b. $\sqrt[5]{a^3} : \sqrt[3]{a^2}.$

206a. $\sqrt[3]{4a^2} : \sqrt[6]{2a^3}.$

206b. $\sqrt[10]{2a^4} : \sqrt[5]{2a^2}.$

207a. $\sqrt{6a^5} : \sqrt[6]{27a^{-9}}.$

207b. $\sqrt[4]{\frac{4}{a^3}} : \sqrt[12]{4a^{-8}}.$

208a. $10a\sqrt{a} : \sqrt[3]{a^2}.$

208b. $3a\sqrt[3]{a} : \sqrt[5]{a^4}.$

209a. $6a^2\sqrt{3a^{-1}b} : 2a^3\sqrt[3]{2ab^{-1}}.$

209b. $2a^3b\sqrt[3]{a^{-2}b^3} : 6ab^2\sqrt{a^3b^{-7}}.$

210a. $5x^2y : \sqrt[3]{25xy^4}.$

210b. $2x^2y^3 : \sqrt[4]{8x^3y^2}.$

211a. $\frac{24a^5b^2}{d^2}\sqrt[5]{\frac{a^2b^7}{c^2}} : \frac{4a^2}{b}\sqrt[3]{\frac{a^4b^7}{cd^5}}.$

211b. $\frac{2a^2b}{c}\sqrt[3]{\frac{a^3b^2}{c^4d^2}} : \frac{4ab^2}{c^2}\sqrt[5]{\frac{a^6d^2}{b^8c^4}}.$

212. $(a^2b + ax^2)\sqrt[3n]{\frac{x}{a^{n-1}c^3}} : ax\sqrt[2n]{\frac{x^4}{a^nc^2}}.$

213. $(x+y) : \frac{1}{3}\sqrt{x^2 - y^2}.$

214. $(x^2 - y^2) : \frac{a}{x}\sqrt[3]{\frac{2a}{(x+y)^4}}.$

215. $(\sqrt[4]{8a^6b^9} - ab\sqrt[6]{8a^4b^5} + ab^2\sqrt[4]{2a^4b}) : \sqrt[4]{2b}.$

216. $(\sqrt[5]{8x^3} - 3\sqrt{3}) : (\sqrt[5]{2x} - \sqrt{3}).$

217. $(2a\sqrt[3]{ax^2} - a\sqrt[6]{ax^5} - ax) : (\sqrt[3]{a^2x} - \sqrt{ax}).$

218. $(x^2\sqrt[4]{27xy^3} + 2xy\sqrt{2xy}) : (\sqrt[4]{3x^3y} + \sqrt{2xy}).$

Raising Roots to a Power and Extracting a Root from Them

Raising polynomial expressions is done by the general rules for raising a polynomial to a power. To raise a root to a power, raise the radicand to that power: the index of the root remains unchanged, and the exponents of the radicand are multiplied by the exponent of the power. A coefficient is raised to the power separately. In extracting a root from a root, the indices of the roots are multiplied and the radicand remains unchanged.

219a. $(\sqrt[4]{a^3})^4$.

219b. $(\sqrt[7]{a^4})^7$.

220a. $(\sqrt[3]{a^2})^2$.

220b. $(\sqrt[4]{a^3})^5$.

221a. $(\sqrt[4]{2x^3})^5$.

221b. $(\sqrt[3]{4x^2})^2$.

222a. $(-a\sqrt[8]{a^2b^3})^7$.

222b. $(-a\sqrt[5]{a^3x})^4$.

223a. $(a^2x\sqrt[3]{3a^2x})^4$.

223b. $(ax^2\sqrt[3]{2ax^2})^2$.

224a. $\left(-2a\sqrt[6]{\frac{3}{a^4}}\right)^4$.

224b. $\left(-\frac{3}{a^2}\sqrt[6]{\frac{2}{a^2}}\right)^3$.

225a. $(\sqrt[5]{(x-y)^2})^4$.

225b. $(\sqrt[3]{(x+y)^2})^5$.

226a. $\left(\frac{\sqrt[4]{a^{-3}b^2}}{a^{-2}b^3}\right)^{-3}$.

226b. $\left(\frac{a^5b^{-3}}{\sqrt[5]{a^{-4}b}}\right)^{-4}$.

227a. $(a^{-1}b^{-2}\sqrt[3]{4a^nb^2})^{-2}$.

227b. $(a^2b^{-1}\sqrt{2a^{-3}b^n})^{-3}$.

228a. $(\sqrt[n]{(x^2+y^2)^m})^{np}$.

228b. $(\sqrt[n]{(x^2-y^2)^n})^{mp}$.

229a. $(\sqrt{3}-\sqrt{2})^2$.

229b. $(\sqrt{5}+\sqrt{2})^2$.

230a. $\left(\frac{1}{2}+2\sqrt{2}\right)^2$.

230b. $\left(2\sqrt{3}-\frac{1}{3}\right)^2$.

231a. $(\sqrt[3]{4}+\sqrt{2})^2$.

231b. $(\sqrt[6]{2}-\sqrt{3})^2$.

232a. $(\sqrt{3}-2\sqrt[3]{2})^3$.

232b. $(\sqrt{2}+3\sqrt[3]{3})^3$.

233a. $(\sqrt{2}-\sqrt{3}+\sqrt{6})^2$.

233b. $(\sqrt{2}+\sqrt{5}-\sqrt{10})^2$.

234a. $(3\sqrt{2}-2\sqrt{5}-\sqrt{10})^2$.

234b. $(5\sqrt{6}+3\sqrt{2}-2\sqrt{3})^2$.

235a. $(\sqrt{3+\sqrt{5}}+\sqrt{3-\sqrt{5}})^2$.

235b. $(\sqrt{7+2\sqrt{6}}+\sqrt{7-2\sqrt{6}})^2$.

236a. $(\sqrt{11+6\sqrt{2}}-\sqrt{11-6\sqrt{2}})^2$.

236b. $(\sqrt{13+5\sqrt{4.2}}+\sqrt{13-5\sqrt{4.2}})^2$.

237a. $(\sqrt{11+4\sqrt{7}}-\sqrt{11-4\sqrt{7}})^2$.

237b. $(\sqrt{\sqrt{7}-\sqrt{3}}-\sqrt{\sqrt{7}+\sqrt{3}})^2$.

238a. $\left(\frac{b}{4}\sqrt{ab}-\frac{2}{\sqrt{a}}\right)^2$.

238b. $\left(\frac{a}{2}\sqrt{\frac{c}{b}}-\frac{3}{\sqrt{ab}}\right)^2$.

239a. $(a\sqrt{a}+a\sqrt{2a})^3$.

239b. $(a\sqrt{b}-2a\sqrt{2b})^3$.

If the given root has a coefficient, it is usually brought under the radical of that root before a further root is extracted.

Extract the root:

240a. $\sqrt{\sqrt[3]{a^2}}$.

240b. $\sqrt[3]{\sqrt[4]{a^3}}$.

241a. $\sqrt[3]{\sqrt[5]{a^4}}$.

241b. $\sqrt{\sqrt[5]{a^3}}$.

242a. $\sqrt[3]{\sqrt{125}}$.

242b. $\sqrt[4]{\sqrt[5]{81}}$.

243a. $\sqrt[4]{\sqrt{256a^{10}}}$.

243b. $\sqrt[9]{\sqrt[3]{512a^{18}}}$.

244a. $\sqrt{a\sqrt[4]{a^3}}$.

244b. $\sqrt[3]{a\sqrt[3]{a}}$.

245a. $\sqrt[4]{a^2\sqrt[3]{a^4}}$.

245b. $\sqrt[6]{a^3\sqrt[5]{a^6}}$.

246a. $\sqrt{\sqrt[4]{a^{10}b^2c^8}}$.

246b. $\sqrt[3]{\sqrt[10]{a^{10}b^5c^{15}}}$.

247a. $\sqrt{\sqrt[3]{a^2}\cdot\sqrt{b}}$.

247b. $\sqrt[3]{\sqrt[5]{a^4}\cdot\sqrt{\sqrt[3]{a^2}}}$.

248a. $\sqrt{x^3\sqrt[3]{x\sqrt[4]{x}}}$.

248b. $\sqrt{x^2\sqrt[3]{x\sqrt{x}}}$.

249a. $\sqrt{x\sqrt[3]{\frac{x^2}{y}\sqrt{\frac{y}{x}}}}$.

249b. $\sqrt[3]{\frac{x}{y}\sqrt{\frac{x}{y}\sqrt{\frac{y^3}{x}}}}$.

250. $\sqrt[4]{2x\sqrt[3]{2x^2y\cdot 3y\sqrt{3xy^3}}}$.

251. $\sqrt[4]{20736}$.

252. $\sqrt[10]{59049}$.

253. $\sqrt[12]{4096}$.

254. $\sqrt[9]{262144}$.

Removing Irrationality in the Denominator of a Fraction

To remove irrationality in the denominator, find the simplest expression which, multiplied by the denominator, gives a rational expression, and multiply both numerator and denominator by it. For example, if the denominator is a binomial $\sqrt{a}\pm\sqrt{b}$, multiply numerator and denominator by the conjugate quantity (sum by the corresponding difference and conversely). In more complex cases the irrationality is removed in several steps, successively introducing factors into the terms of the fraction.

Remove the irrationality in the denominator of the fraction:

255a. $\frac{a}{\sqrt{a}}$.

255b. $\frac{b^2}{\sqrt{b}}$.

256a. $\frac{m}{\sqrt{m^3}}$.

256b. $\frac{n}{\sqrt{n^5}}$.

257a. $\frac{a}{\sqrt[3]{a^4}}$.

257b. $\frac{a}{\sqrt[5]{a^4}}$.

258a. $\frac{m+n}{\sqrt{m-n}}$.

258b. $\frac{m-n}{\sqrt{m+n}}$.

259a. $\frac{4}{\sqrt{2}}$.

259b. $\frac{15}{\sqrt{5}}$.

260a. $\frac{6}{\sqrt[4]{8}}$.

260b. $\frac{6}{\sqrt[4]{12}}$.

261a. $\frac{\sqrt[6]{49}}{\sqrt[3]{21}}$.

261b. $\frac{\sqrt[8]{36}}{\sqrt[4]{12}}$.

262. $\frac{a^2-b^2}{\sqrt[3]{a-b}}$.

263. $\frac{a-b}{\sqrt[3]{a^2-b^2}}$.

264a. $\frac{\sqrt{a}-\sqrt{b}}{\sqrt{a}+\sqrt{b}}$.

264b. $\frac{\sqrt{a}+\sqrt{b}}{\sqrt{a}-\sqrt{b}}$.

265a. $\frac{a}{1-\sqrt{a}}$.

265b. $\frac{a}{\sqrt{a}+1}$.

266a. $\frac{\sqrt{3}}{2+\sqrt{3}}$.

266b. $\frac{\sqrt{2}}{3-\sqrt{2}}$.

267a. $\frac{1-a}{\sqrt{1-\sqrt{a}}}$.

267b. $\frac{1-a}{\sqrt{1+\sqrt{a}}}$.

268a. $\frac{n}{\sqrt[3]{a}-\sqrt[3]{b}}$.

268b. $\frac{n}{\sqrt[3]{a}+\sqrt[3]{b}}$.

269. $\frac{12}{3+\sqrt{2}-\sqrt{3}}$.

270. $\frac{2+\sqrt{30}}{\sqrt{5}+\sqrt{6}-\sqrt{7}}$.

271. $\frac{1+3\sqrt{2}-2\sqrt{3}}{\sqrt{2}+\sqrt{3}+\sqrt{6}}$.

272. $\frac{n}{\sqrt{\sqrt{2}+\sqrt[8]{3}}}$.

273. $\frac{\sqrt{a^2-b^2}+\sqrt{a^2+b^2}}{\sqrt{a^2-b^2}-\sqrt{a^2+b^2}}$.

274. $\frac{1}{\sqrt{2}+\sqrt[4]{3}}$.

275. $\frac{47}{2\sqrt{3}-\sqrt[4]{3}}$.

Square Root of a Binomial of the Form $A \pm \sqrt{B}$

The square root of a binomial $A \pm \sqrt{B}$ can be simplified when $A^2 - B$ is a perfect square. The formula is

$$\sqrt{A \pm \sqrt{B}} = \sqrt{\frac{A + \sqrt{A^2 - B}}{2}} \pm \sqrt{\frac{A - \sqrt{A^2 - B}}{2}}.$$

By this formula solve the following examples:

276a. $\sqrt{2+\sqrt{3}}$.

276b. $\sqrt{4-\sqrt{7}}$.

277a. $\sqrt{6+4\sqrt{2}}$.

277b. $\sqrt{7+2\sqrt{10}}$.

278a. $\sqrt{5-\sqrt{21}}$.

278b. $\sqrt{8-\sqrt{15}}$.

279a. $\sqrt{7+4\sqrt{3}}$.

279b. $\sqrt{11+4\sqrt{7}}$.

280a. $\sqrt{4\sqrt{2}+2\sqrt{6}}$.

280b. $\sqrt{5\sqrt{5}-2\sqrt{30}}$.

281a. $\sqrt{4\sqrt{5}-2\sqrt{15}}$.

281b. $\sqrt{3\sqrt{7}+2\sqrt{14}}$.

282a. $\sqrt{\sqrt{14+6\sqrt{5}}}$.

282b. $\sqrt{\sqrt{124-32\sqrt{15}}}$.

283. $\sqrt{17+6\sqrt{4-\sqrt{9+4\sqrt{2}}}}$.

284. $\sqrt{(a+b)-2\sqrt{ab}}.$

285. $\sqrt{2a^2+2\sqrt{a^4-b^2}}.$

286. $\sqrt{\frac{19}{20}+\sqrt{\frac{3}{5}}}.$

287. $\sqrt{0.33+3\sqrt{0.0091}}.$

288. $\sqrt{3a+2a\sqrt{2}}+\sqrt{3a-2a\sqrt{2}}.$

Problems on All Operations with Radicals

289a. $\frac{3}{5-\sqrt{5}}-\frac{1}{3+\sqrt{5}}.$

289b. $\frac{1}{5+\sqrt{5}}+\frac{1}{3-\sqrt{5}}.$

290a. $\frac{5}{4-\sqrt{11}}-\frac{4}{\sqrt{11}-\sqrt{7}}-\frac{2}{3+\sqrt{7}}.$

290b. $\frac{9}{5-\sqrt{7}}+\frac{22}{7+\sqrt{5}}-\frac{1}{\sqrt{7}+\sqrt{5}}.$

291a. $\frac{1}{7+4\sqrt{3}}+\frac{1}{7-4\sqrt{3}}.$

291b. $\frac{\sqrt{5}-1}{\sqrt{5}+1}+\frac{\sqrt{5}+1}{\sqrt{5}-1}.$

292a. $\frac{\sqrt{12}}{(1+\sqrt{2})(\sqrt{6}-\sqrt{3})}.$

292b. $\frac{1}{2+\sqrt{3}}+2+\sqrt{3}.$

293a. $\frac{3-\sqrt{5}}{(\sqrt{3}+\sqrt{5})^2}+\frac{3+\sqrt{5}}{(\sqrt{3}-\sqrt{5})^2}.$

293b. $\frac{4}{3\sqrt{5}-5}-\frac{3}{\sqrt{5}}.$

294a. $\frac{1}{5+\sqrt{5}}+\frac{1}{7-\sqrt{29}}.$

294b. $\frac{7}{4-\sqrt{7}}-\frac{16}{7+\sqrt{13}}.$

295a.
$\frac{5}{4-\sqrt{11}}+\frac{1}{3+\sqrt{7}}-\frac{6}{\sqrt{7}-2}-\frac{1}{2}(\sqrt{7}-5).$

295b. $\frac{\sqrt{a}}{\sqrt{a}-\sqrt{x}}-\frac{\sqrt{x}}{\sqrt{a}+\sqrt{x}}.$

296. $\frac{\sqrt{2}-1}{\sqrt{2}+1}+\frac{\sqrt{2}+1}{\sqrt{2}-1}.$

297. $\frac{14}{\sqrt{\frac{2}{3}x-4}}-\frac{3}{\sqrt{\frac{3}{2}x-9}}.$

298. $a\sqrt{\frac{a+b}{a-b}}-b\sqrt{\frac{a-b}{a+b}}-\frac{2b^2}{\sqrt{a^2-b^2}}.$

299. $\left(\frac{1}{\sqrt{1+x}}+\sqrt{1-x}\right)\cdot\left(\frac{1}{\sqrt{1-x^2}}+1\right).$

300. $\frac{a+\sqrt{a^2-x^2}}{a-\sqrt{a^2-x^2}}-\frac{a-\sqrt{a^2-x^2}}{a+\sqrt{a^2-x^2}}.$

301. $\frac{17}{2}\sqrt{4\sqrt{63}}-10\sqrt{3\sqrt{7}}-3\sqrt{6\sqrt{28}}.$

302. $x\sqrt{x\sqrt{x\sqrt{x}}}-2\sqrt{x^3\sqrt[4]{x^3}}+3\sqrt[-2]{x^{-5}\sqrt[4]{x^5}}-4x^2\sqrt[-4]{x\sqrt{\frac{1}{x}}}.$

303. $\frac{a+2+\sqrt{a^2-4}}{a+2-\sqrt{a^2-4}}+\frac{a+2-\sqrt{a^2-4}}{a+2+\sqrt{a^2-4}}.$

304. $\left[\frac{a}{2\sqrt{a}}\cdot\frac{x}{2\sqrt{x}}:\left(\sqrt[3]{\frac{a^2}{x}}\cdot a^{-1}\sqrt{x}\right)^6\right]\cdot\sqrt[-2]{\frac{1}{4}a^2\sqrt{\frac{a}{x}}}.$

305. $\left(1+\sqrt{\frac{a-x}{a+x}}\right):\left(1-\sqrt{\frac{a-x}{a+x}}\right).$

306. $5a\sqrt{a\sqrt{a\sqrt{a}}}-2\sqrt{a^3\sqrt[4]{a^3}}+3\sqrt[-2]{a^{-5}\sqrt[4]{a^5}}-4a^2\sqrt[-4]{a\sqrt{\frac{1}{a}}}.$

307. $\left(-4a\sqrt[3]{a^{-2}\sqrt{ax}}\right)^3+\left(-10a\sqrt{x}\sqrt[4]{\frac{1}{ax}}\right)^2-5\left[\left(\sqrt[3]{a\sqrt[-4]{\frac{a}{x}}}\right)^3\right]^2.$

308. $\sqrt[-2]{\frac{1}{4}a^2\sqrt{\frac{a}{x}}}\cdot\left[\frac{a}{2\sqrt{a}}\cdot\frac{a}{2\sqrt{x}}:\left(\sqrt[3]{\frac{a^2}{x}}\cdot a^{-1}\sqrt{x}\right)^6\right].$

309. $\left[\sqrt{\frac{(1-a)\sqrt[3]{1+a}}{a}}\cdot\sqrt[3]{\frac{3a^2}{4-8a+4a^2}}\right]^{-1}:\sqrt[3]{\frac{3a\sqrt{a}}{2\sqrt{1-a^2}}}.$

310. $\left(\frac{1}{\sqrt{1+x}}+\sqrt{1-x}\right):\left(\sqrt{1-x^2}+1\right).$

311. $\left(\frac{ax+n^2}{\sqrt{a^2nx-an^3}}-\sqrt{\frac{n}{x}}\cdot\frac{2nx}{\sqrt{ax-n^2}\cdot\sqrt{ax}}\right):\sqrt[4]{\frac{x}{an^2}}-a^{-2}.$

312. $\sqrt{\frac{a}{b}\left(\frac{c}{a}+\frac{b}{c}\right)+2\sqrt{\frac{a}{b}}}\cdot\sqrt{\frac{a}{b}\left(\frac{c}{a}+\frac{b}{c}\right)-2\sqrt{\frac{a}{b}}}.$

313. $\frac{2+\sqrt{3}}{\sqrt{2}+\sqrt{2+\sqrt{3}}}+\frac{2-\sqrt{3}}{\sqrt{2}-\sqrt{2-\sqrt{3}}}.$

Determine the particular values of the expressions:

314. $\frac{1+x}{1+\sqrt{1+x}}+\frac{1-x}{1-\sqrt{1-x}}$ for $x=\frac{\sqrt{3}}{2}$.

315. $\frac{x+1}{x+\sqrt{x^2+x}}+\frac{x-1}{x-\sqrt{x^2-x}}$ for $x=\frac{2}{\sqrt{3}}$.

316. $\frac{2a\sqrt{1+x^2}}{x+\sqrt{1+x^2}}$ for $x=\frac{1}{2}\left(\sqrt{\frac{a}{b}}-\sqrt{\frac{b}{a}}\right)$.

317. $\frac{1-ax}{1+ax}\sqrt{\frac{1+bx}{1-bx}}$ for $x=\frac{1}{a}\sqrt{\frac{2a}{b}-1}$.

Powers and Roots with Fractional Exponents

A quantity with a fractional exponent represents a root whose index equals the denominator of the fraction, of the same quantity raised to the power given by the numerator. For example, $a^{\frac{3}{5}}=\sqrt[5]{a^3}$; in general, $a^{\frac{m}{n}}=\sqrt[n]{a^m}$. A root with a fractional index equals a power whose exponent is the reciprocal of the index of the root. For example, $\sqrt[\frac{3}{2}]{a}=a^{\frac{2}{3}}$; in general, $\sqrt[\frac{m}{n}]{a}=a^{\frac{n}{m}}$. Operations with powers and roots having fractional exponents are performed by the same rules as for integer exponents.

Replace the radicals by expressions with fractional exponents:

318a. $\sqrt[3]{a^2}$.

318b. $\sqrt[5]{a^3}$.

319a. $\sqrt[4]{a^{-3}}$.

319b. $\sqrt[3]{a^{-2}}$.

320a. $\sqrt[5]{a^{-3}b^4}$.

320b. $\sqrt[4]{a^3b^{-2}}$.

321a. $\sqrt{a^{-3}}$.

321b. $\sqrt[3]{a^{-5}}$.

322a. $\sqrt{a^2+b^2}$.

322b. $\sqrt[3]{a^3-b^3}$.

323a. $\sqrt[-3]{\frac{a^3-b^3}{a^{-1}b^2}}$.

323b. $\sqrt[-2]{\frac{a^2b^{-2}}{a^2-b^2}}$.

324. $\sqrt[m]{x^{n+1}}$.

325. $\sqrt[mp+mq]{a^{p+q}}$.

326. $\sqrt[m]{\frac{x^2}{y^n}}$.

327. $\sqrt[pr]{\frac{1}{a^{pn}}}$.

328a. $\sqrt[3]{a\sqrt[5]{b}}$. **328b.** $\sqrt[n]{x^2\sqrt{y}}$.

329a. $\sqrt[3]{a^2+x^2-2ax}$. **329b.** $\sqrt[n]{a^2+2ab+b^2}$.

Replace the expressions with fractional exponents by radicals:

330a. $a^{\frac{5}{6}}$. **330b.** $a^{\frac{2}{3}}$.

331a. $a^{-\frac{3}{4}}$. **331b.** $a^{-\frac{3}{7}}$.

332a. $(a+b)^{\frac{2}{3}}$. **332b.** $(a-b)^{\frac{5}{3}}$.

333a. $3a^{\frac{1}{2}}(a-b)^{\frac{3}{5}}$. **333b.** $4a^{-\frac{2}{3}}(a+b)^{-\frac{1}{2}}$.

334a. $(4a)^{-\frac{1}{4}}$. **334b.** $a^{\frac{k-1}{k}}$.

335a. $(a+b)^{0.25}$. **335b.** $(a-x)^{-1.5}$.

336a. $m^{-\frac{1}{2}x}$. **336b.** $\left(\frac{1}{x}\right)^{-0.3n}$.

337a. $x^{\frac{1}{4}-\frac{1}{6}n}$. **337b.** $x^{p-\frac{p^2}{p-n}}$.

338a. $e^{-\frac{1}{x}}$. **338b.** $e^{-\frac{x^2}{2}}$.

339a. $n^{2.5}$. **339b.** $\frac{1}{(a-b)^{-0.6}}$.

Compute:

340a. $4^{\frac{1}{2}}$. **340b.** $27^{\frac{1}{3}}$.

341a. $81^{\frac{3}{4}}$. **341b.** $16^{\frac{5}{4}}$.

342a. $16^{-\frac{5}{4}}$. **342b.** $32^{-\frac{4}{5}}$.

343a. $(-8)^{\frac{2}{3}}$. **343b.** $(-27)^{\frac{4}{3}}$.

344a. $\left(\frac{25}{36}\right)^{-\frac{1}{2}}$. **344b.** $\left(\frac{27}{8}\right)^{-\frac{1}{3}}$.

345a. $\left(-3\frac{3}{8}\right)^{-\frac{2}{3}}$. **345b.** $\left(-1\frac{61}{64}\right)^{-\frac{2}{3}}$.

346a. $(0.64)^{0.5}$. **346b.** $(0.027)^{\frac{2}{3}}$.

347a. $81^{-0.75}$. **347b.** $1024^{-0.6}$.

348a. $8^{\frac{2}{3}}-16^{\frac{1}{4}}+9^{\frac{1}{2}}$. **348b.** $25^{\frac{1}{2}}-27^{\frac{2}{3}}+81^{\frac{3}{4}}$.

349a. $16^{0.5}+\left(\frac{1}{16}\right)^{-0.75}-\left(\frac{1}{2}\right)^{-6}$. **349b.** $9^{-0.5}-8^{-1\frac{1}{3}}+(0.25)^{-\frac{3}{2}}$.

Carry out the indicated operations:

350a. $a^{\frac{2}{3}}b^{\frac{3}{5}}\cdot a^{\frac{3}{4}}b^{\frac{2}{3}}$. **350b.** $a^{\frac{5}{2}}b^{\frac{2}{3}}\cdot a^{\frac{2}{3}}b^{\frac{3}{4}}$.

351a. $a^{\frac{7}{12}}b^{\frac{5}{6}}:a^{\frac{2}{3}}b^{\frac{3}{4}}$. **351b.** $a^{\frac{11}{15}}b^{\frac{2}{3}}:a^{\frac{3}{5}}b^{\frac{5}{6}}$.

352a. $\left(a^{\frac{3}{2}}-b^{\frac{3}{2}}\right):\left(a^{\frac{1}{2}}-b^{\frac{1}{2}}\right)$. **352b.** $\left(a^{\frac{3}{2}}+b^{\frac{3}{2}}\right):\left(a^{\frac{1}{2}}+b^{\frac{1}{2}}\right)$.

353. $\left(a^{\frac{1}{2}}+a^{\frac{1}{4}}b^{\frac{1}{4}}+b^{\frac{1}{2}}\right)\cdot\left(a^{\frac{1}{2}}-a^{\frac{1}{4}}b^{\frac{1}{4}}+b^{\frac{1}{2}}\right)$.

354. $\left(a^{\frac{3n}{2}}+b^{-\frac{3n}{2}}\right):\left(a^{\frac{n}{2}}+b^{-\frac{n}{2}}\right)$.

355. $\left(a^{\frac{4}{3}}+4a^{\frac{2}{3}}b^{\frac{2}{3}}+16b^{\frac{4}{3}}\right):\left(a^{\frac{2}{3}}+2a^{\frac{1}{3}}b^{\frac{1}{3}}+4b^{\frac{2}{3}}\right)$.

356. $\left(a^{\frac{1}{2}}-b^{\frac{1}{2}}-c^{\frac{1}{2}}+2b^{\frac{1}{4}}c^{\frac{1}{4}}\right):\left(a^{\frac{1}{4}}+b^{\frac{1}{4}}-c^{\frac{1}{4}}\right)$.

357a. $\left(a^{\frac{3}{2}} - a^{\frac{1}{6}}b^{\frac{1}{2}} + b^{\frac{3}{2}}\right)^2.$

357b. $\left(a^{\frac{3}{4}} + a^{\frac{1}{4}}b^{\frac{1}{2}} - b^{\frac{5}{4}}\right)^2.$

358a. $\left(a^{\frac{1}{2}}b^{\frac{2}{3}} - 2a^{\frac{2}{3}}b^{\frac{1}{4}}\right)^3.$

358b. $\left(a^{\frac{2}{3}}b^{\frac{1}{2}} - 3a^{\frac{3}{4}}b^{\frac{2}{3}}\right)^3.$

359a.
$\left[\left(a^{-\frac{3}{2}}b\right) \cdot (ab^{-2})^{-\frac{1}{2}} \cdot (a^{-1})^{-\frac{2}{3}}\right]^3.$

359b.
$\left[\left(a^{\frac{2}{3}}b^{-1}\right)^2 \cdot (a^2b^{-1})^{\frac{1}{2}} \cdot \left(b^{\frac{2}{3}}\right)^{-\frac{3}{2}}\right]^2.$

360. $\frac{a-b}{a^{\frac{1}{2}}-b^{\frac{1}{2}}} - \frac{a^{\frac{3}{2}}-b^{\frac{3}{2}}}{a-b}.$

361. $\sqrt{a^{\frac{3}{2}}b^{-2} - 6a^{\frac{3}{4}}b^{-\frac{1}{3}} + 9b^{\frac{4}{3}}}.$

362. $\sqrt[0.4]{\frac{a-2b^3\sqrt{2a^6b^{-3}}}{\left(\sqrt{a-5b^3}\right)^{\frac{4}{15}}}}.$

363. $\sqrt[\frac{2}{3}]{a^{\frac{4}{3}} + a - 2a^{\frac{7}{6}}}.$

364. $\left(a^{\frac{1}{4}} + b^{\frac{1}{4}}\right) : \left(\sqrt[3]{\frac{a\sqrt[3]{b}}{b\sqrt{a^3}}} + \sqrt[\frac{1}{2}]{\frac{\sqrt{a}}{a\sqrt[8]{b^3}}}\right).$

Imaginary Numbers

Even-degree roots of negative numbers (rational and irrational) are called *imaginary*; positive and negative numbers are called *real.* The simplest imaginary number is $\sqrt{-1}$, denoted i, so that $\sqrt{-1} = i$ and $i^2 = -1$. Raising i to successive powers: $i^1 = i$; $i^2 = -1$; $i^3 = -i$; $i^4 = +1$; these four results recur periodically, so any positive integer power of i equals the power whose exponent is the remainder of dividing the given exponent by 4. For example:

$$i^{26} = i^2 = -1; \qquad i^{35} = i^3 = -i.$$

Every imaginary number of the form $\sqrt{-a}$ can be represented as the product of a real quantity by i, namely:

$$\sqrt{-a} = i\sqrt{a}.$$

Such an expression for an imaginary number is called its *normal form.* To carry out operations with imaginary numbers one must first bring them to normal form.

An expression of the form $a + bi$, where a and b are real numbers, is called a *complex number.* It becomes real in the case $b = 0$. Two complex numbers of the form $a + bi$ and $a - bi$, that is, ones which differ only in the signs at the imaginary part, are called *conjugate.* In the theory of operations with complex numbers the quantity $\sqrt{a^2 + b^2}$ often occurs. It is called the *modulus* of the complex number $a + bi$ and is usually denoted by M.

In carrying out any operations with complex numbers one must first bring their imaginary parts to normal form.

In the addition and subtraction of complex numbers their real parts and their imaginary parts are added or subtracted separately. Thus:

$$a + bi \pm (a_1 + b_1 i) = (a \pm a_1) + (b \pm b_1)i.$$

Multiplication of complex numbers is performed by the general rules, only taking into account that $i^2 = -1$. Therefore:

$$(a+bi)(a_1+b_1i) = aa_1 + a_1bi + ab_1i - bb_1 = (aa_1 - bb_1) + (a_1b + ab_1)i.$$

Division of complex numbers is carried out by multiplying the dividend and the divisor by the expression conjugate to the divisor. By this the new divisor becomes real; namely, it turns into the square of the modulus of the former divisor. Thus:

$$(a+bi):(a_1+b_1i) = \frac{(a+bi)(a_1-b_1i)}{a_1^2+b_1^2} = \frac{aa_1+bb_1}{M_1^2} + \frac{a_1b-ab_1}{M_1^2}i.$$

Raising complex numbers to the square and to the cube is done by the formulas for real numbers. In applying these formulas it is useful at first to denote only the powers of the imaginary i, and afterwards to replace them by the simplest expressions. Thus:

$$(a+bi)^3 = a^3 + 3a^2bi + 3ab^2i^2 + b^3i^3 = a^3 - 3ab^2 + (3a^2b - b^3)i.$$

Extraction of the square root is done by the formulas

$$\sqrt{a \pm bi} = \sqrt{\frac{M+a}{2}} \pm \sqrt{\frac{M-a}{2}}\,i,$$

where M denotes the modulus of the complex number under the radical. To the root obtained one may assign either those signs of its real and imaginary parts which they have by this formula, or the opposite signs.

365. $(\sqrt{-1})^6$.

366. $(\sqrt{-1})^{21}$.

367. $(\sqrt{-1})^7$.

368. $(\sqrt{-1})^{56}$.

369. i^{40}.

370. i^{37}.

371. i^{18}.

372. i^{4n+2}.

373a. i^{4n+3}. **373b.** i^{4n-1}.

374. i^{8n+5}.

Simplify the imaginary expressions:

375. $\sqrt{-4}$.

376. $\sqrt{-81}$.

377. $\sqrt{-a^2}$.

378. $\sqrt{-b}$.

379. $\sqrt{-\frac{9}{4}}$.

380. $\sqrt{-\frac{a^4}{b^8}}$.

381. $\sqrt{-a}$.

382. $\sqrt{-9x}$.

383a. $\sqrt{-a^2-b^2}$.

383b. $\sqrt{-(a-b)^2}$.

384. $\sqrt{-x^2-y^2+2xy}$.

Carry out the indicated operations:

385a.
$\sqrt{-25}+\sqrt{-49}-\sqrt{-64}+\sqrt{-1}$.

385b.
$\sqrt{-144}-\sqrt{-81}-\sqrt{-1}+\sqrt{-9}$.

386a.
$3\sqrt{-4}+5\sqrt{-27}-3\sqrt{-16}-5\sqrt{-3}$.

386b.
$10\sqrt{-25}-5\sqrt{-8}+\sqrt{-49}-2\sqrt{-2}$.

387. $3+2i+(4-3i)-[(8-5i)-(5+13i)]$.

388. $a+bi-(2a-3bi)+[(a-4bi)+(5a-2bi)]$.

389a. $\sqrt{-16}\cdot\sqrt{-9}$.

389b. $\sqrt{-8}\cdot\sqrt{-2}$.

390a. $\sqrt{-a}\cdot\sqrt{-b}$.

390b. $\sqrt{-m}\cdot\sqrt{-n}$.

391a. $i\sqrt{-x^2}$.

391b. $-i\sqrt{-y^2}$.

392. $\sqrt{a-b}\cdot\sqrt{b-a}$.

393. $(2-5i)(8-3i)$.

394. $(5+2\sqrt{-7})\cdot(6-5\sqrt{-7})$.

395. $(\sqrt{a}-\sqrt{-b})\cdot(\sqrt{a}+3\sqrt{-b})$.

396. $(3\sqrt{-5}-2\sqrt{-7})\cdot(2\sqrt{-7}+3\sqrt{-5})$.

397a. $a:\sqrt{-a}$.

397b. $ai:\sqrt{-a}$.

398. $\sqrt{-ax}:\sqrt{-x}$.

399a. $\frac{a^2+b^2}{a-bi}$.

399b. $\frac{a^2+b^2}{a+bi}$.

400a. $\frac{x-y}{x+yi}$.

400b. $\frac{x-y}{x-yi}$.

401. $\frac{4}{1+\sqrt{-3}}$.

402. $\frac{3-5i\sqrt{8}}{3+5i\sqrt{8}}$.

403a. $\frac{36-\sqrt{-2}}{2+3i\sqrt{2}}$.

403b. $\frac{5-29i\sqrt{5}}{7-3\sqrt{-5}}$.

404. $\frac{2-\sqrt{-7}}{3+\sqrt{-21}}$.

405a. $(a+bi)^2$.

405b. $(a-bi)^2$.

406. $(3-\sqrt{-2})^2$.

407. $\left(\frac{1+\sqrt{-3}}{2}\right)^2$.

408. $(3\sqrt{-5}+2\sqrt{-1})^2$.

409. $(2-3\sqrt{-2})^2$.

410. $\left(\frac{-1+2\sqrt{-2}}{2}\right)^2$.

411a. $(a-bi)^3$.

411b. $(a+bi)^3$.

412. $(3+\sqrt{-2})^3$.

413. $(\sqrt{-3} - 2\sqrt{-1})^3$.

414. $\left(\frac{-1+\sqrt{-3}}{2}\right)^3$.

415. $\sqrt{3 + 4\sqrt{-1}}$.

416. $\sqrt{-3 - 4i}$.

417. $\sqrt{1 + 4\sqrt{-3}}$.

418. $\sqrt{2 - 3\sqrt{-5}}$.

419. $\sqrt{20 - 4\sqrt{-11}}$.

420. $\sqrt{6 + \sqrt{-13}}$.

421a. $\sqrt{\sqrt{-1}}$.

421b. $\sqrt{-\sqrt{-1}}$.

422. $\sqrt[8]{-1}$.

Chapter X

Functions and Their Graphs

1. Plot the points with coordinates: $(5,\ 8)$; $(-9,\ 3)$; $(-6, -10)$; $(12, -7)$.

2. Plot the points with coordinates: $\left(10\frac{1}{2},\ 3\frac{1}{2}\right)$; $\left(-18,\ 7\frac{1}{2}\right)$; $\left(-8\frac{1}{2},\ -8\frac{1}{2}\right)$; $(2.7, -3.6)$.

3. What features are there in the arrangement of the points $(12,\ 0)$; $(-7,\ 0)$; $(0,\ 0)$; $(0,\ 9)$; $(0, -17)$; $(0, -8.6)$?

4. On what line do the points lie whose coordinates are equal both in magnitude and in sign? How is this line situated? What angles does it form with the coordinate axes?

5. On what line do the points lie whose coordinates are equal in absolute value but opposite in sign? What angles does it form with the coordinate axes?

6. Construct the straight-line segment joining the two points: $(10,\ 13)$ and $(-14, -3)$; $(-5,\ 7)$ and $(3, -9)$.

7. Construct a triangle from the coordinates of its vertices: $(6,\ 12)$; $(9,\ 1)$; $(-6,\ 5)$.

8. Construct a quadrilateral from the coordinates of its vertices: $(-4,\ 9)$; $(12,\ 7)$; $(3, -3)$; $(-10, -6)$.

9. Plot the graphs: $y = 3x$; $2x = 5y$; $3x - 2y = 0$; $4x - 7y = 0$.

10. Plot the graphs: $x + y = 2$; $x - y = 1$.

11. Construct the line $y = 2x + 5$ and find its point of intersection with the axis of ordinates.

12. Construct the line $y = 3x - 7$ and find its point of intersection with the axis of abscissas.

13. Construct the line $y = -x + 3$ and find its points of intersection with the coordinate axes.

14. Construct the line $y = x - 2$ and find its points of intersection with the coordinate axes.

15. Plot the graphs: $xy = 8$; $xy = 1$.

16. Plot the graphs: $xy = -5$; $xy = -9$.

17. Plot a graph for converting inches into centimeters, knowing that 1 inch = 2.54 cm.

18. Plot the graph of the dependence of the circumference of a circle on its radius (take $\pi = 3.14$).

19. The average speed of a train is 35 km per hour. Compile a table of numerical values of the distance travelled by the train over various time intervals and plot the graph of this dependence (hint: take the scale unit on the Y-axis equal to $\frac{1}{10}$ of the scale unit on the X-axis).

20. Plot the graph of the equation of uniform motion $s = s_0 + vt$, taking $s_0 = 2$ cm and $v = 3$ cm/sec.

21. Represent graphically the formula

$$\lambda = 606.5 + 0.305t,$$

expressing the dependence of the heat quantity of water vapor λ on the temperature t.

22. By Faraday's law, the amount Q of substance deposited by the passage of a current is directly proportional to the electrochemical equivalent α, the current strength I, and the time t. Plot the graph of this dependence for copper ($\alpha = 0.328$) at a constant current $I = 24$.

23. Experiment gave the following table for the speed of sound v (m/sec) in dry air at various temperatures t (°C):

t (°C)	−30	−17	−5	0	8	12	20	30
v (m/sec)	313	321	329	332	337	339	344	349

Form an approximate linear formula for the dependence of v on t.

24. Testing an electric locomotive gave the following table of the relation between the current I (amp) and the traction P (kg):

I (amp)	65	86	106	116	137	150
P (kg)	160	360	560	660	850	980

Draw the graph of the dependence of P on I; replace this graph approximately by a straight line; form an approximate linear formula for the dependence of P on I, finding the slope and the initial ordinate by measurement on the drawing.

25. By Ohm's law $I = \frac{V}{R}$, where I (amp) is the current strength, V (volt) the voltage, and R (s/cm) the resistance of the conductor. Taking R constant, plot the graph of I against V; taking V constant, plot the graph of I against R.

26. At a temperature of 14°C and a pressure of 5 atmospheres a gas occupies a volume of 6 L. On the basis of the Boyle–Mariotte law, plot the graph of the change in volume of this mass of gas as a function of the change in pressure.

27. Solve graphically the system of equations:

$$\begin{cases} x + y = 7; \\ x - y = 3. \end{cases}$$

28. Solve graphically the system of equations:

$$\begin{cases} y = 2x - 14; \\ y = \frac{2}{3}x - 2. \end{cases}$$

29. Solve graphically the system of equations:

$$\begin{cases} y = -3x - 9; \\ y = \frac{1}{2}x - 2. \end{cases}$$

30. A pedestrian left a town for a village at 12 noon and walked uniformly at a speed of 3 km per hour. At 2 p.m. of the same day another pedestrian set out from the town along the same road, covering $4\frac{1}{2}$ km per hour. When and at what distance from the town will the second pedestrian overtake the first? (Solve graphically.)

31. The cost price K (in kopecks) of 1 m of a welded seam for a sheet of thickness d millimeters can be computed approximately by the formulas

$$K = 1.5 + 0.95d \quad \text{(for oxy-acetylene welding)},$$

$$K = 4 + 0.6d \quad \text{(for water-gas welding)}.$$

From what sheet thickness on will water-gas welding be cheaper than oxy-acetylene?

32. A cyclist left a town at 9 a.m. and rode at a speed of 12 km per hour; at 10:30 an automobile set out along the same route at a speed of 48 km per hour. Fifteen minutes after departure the automobile stopped for 15 minutes, after which it proceeded at a speed of 30 km per hour. Determine graphically the place and time at which the automobile overtakes the cyclist. (Take the scale on the Y-axis equal to $\frac{1}{10}$ of the scale unit on the X-axis.)

33. From Moscow to Kursk, the distance between which is 530 km, an automobile set out at 6 a.m. and travelled the whole distance without stopping at an average speed of 50 km per hour. Another automobile set out along the same route at 4:30 a.m. and travelled at a speed of 65 km per hour. Having covered 162 km, this latter automobile stopped until 8:30, after which it proceeded at a speed of 40 km per hour as far as Oryol (380 km from Moscow). At Oryol the automobile stayed $1\frac{1}{2}$ hours and then proceeded at a speed of 90 km per hour. Draw the graph of the motion of both automobiles. Where and when did the automobiles meet? Which automobile arrived at Kursk earlier, and by how much time? (Take the scale on the Y-axis equal to $\frac{1}{10}$ of the scale unit on the X-axis.)

Chapter XI

Quadratic Equations

Solving Literal Quadratic Equations

The reduction of literal quadratic equations to the simplest form and their solution are carried out by the same methods and the same formulas as were indicated for quadratic equations with numerical coefficients (see Part I of the problem book).

Solve the incomplete quadratic equations:

1a. $ax^2 + bx = 0.$ **1b.** $\frac{x^2}{a} - \frac{x}{b} = 0.$

2. $(x-a)^2 + (x-b)^2 = a^2 + b^2.$

3a. $n^2x^2 + mx = m^2x^2 + nx.$ **3b.** $n^4x^2 - m^2x = m^4x^2 - n^2x.$

4a. $\frac{x+a}{x-a} + \frac{x-a}{x+a} = \frac{a(3x+2a)}{x^2-a^2}.$ **4b.** $\frac{x+b}{x-a} - \frac{x-b}{x+a} = \frac{4x^2}{x^2-a^2}.$

5a. $x^2 = 9a^2b^2.$ **5b.** $x^2 - 16a^4c^2 = 0.$

6a. $\frac{x-a}{a} = \frac{a}{x-a}.$ **6b.** $\frac{2x+a}{a} = \frac{a}{2x+a}.$

7a. $\frac{x+a}{x+b} = \frac{a-x}{x-b}.$ **7b.** $\frac{x-2a}{x+2a} = \frac{b-x}{x+b}.$

8. $1 + 2x + x^2 = n^2\left(1 + \frac{2x}{n^2} + \frac{x^2}{n^4}\right).$

9a. $\frac{ax}{a+1} = \frac{a+1}{ax}.$ **9b.** $\frac{ax-3}{a} = \frac{a+6}{ax+3}.$

10a. $ax^2 - b^3 = a^3 - bx^2.$ **10b.** $a^2x^2 + b^4 = a^4 + b^2x^2.$

11. $\frac{2a+b+x}{x+2a-b} = \frac{x-2a+b}{2a+b-x}.$

12. $\frac{x^2+2ax}{x^3-a^3} + \frac{x}{(x+a)^2-ax} = \frac{1}{x-a}.$

Solve the complete quadratic equations:

13a. $x^2 - 4ax + 3a^2 = 0.$ **13b.** $x^2 + 8ax + 15a^2 = 0.$

14a. $x^2 + 2a^3x - 35a^6 = 0.$ **14b.** $x^2 + 6a^2x - 27a^4 = 0.$

15a. $x^2 - 2ax + a^2 - b^2 = 0.$ **15b.** $x^2 - 2bx - a^2 + b^2 = 0.$

16. $x^2 + 2bx - a^2 + 8ab - 15b^2 = 0.$

17a. $2x^2 - 3ax - 2a^2 = 0.$ **17b.** $4x^2 - 20ax + 9a^2 = 0.$

18a. $6x^2 + 5ax + a^2 = 0.$ **18b.** $8x^2 + 2ax - 3a^2 = 0.$

19a. $3b^2x^2 + 10abx + 3a^2 = 0.$ **19b.** $6b^2x^2 - 5abx - 6a^2 = 0.$

20a. $20b^2x^2 - 9abx - 20a^2 = 0.$

20b. $24b^2x^2 + 14abx - 3a^2 = 0.$

21a. $(mx+n)(nx-m) = 0.$

21b. $(n-mx)(nx+m) = 0.$

22a. $ab(x^2+1) - (a^2+b^2)x = 0.$

22b. $ax(bx-a) - c(a-bx) = 0.$

23a. $bx^2 - a = (a-b)x.$

23b. $(a-b)x^2 + 2b = (a+b)x.$

24a. $(a^2-b^2)x^2 + ab = (a^2+b^2)x.$

24b. $(a^2-b^2)x^2 - ab = (a^2+b^2)x.$

25a. $x - \frac{1}{x} = \frac{a}{b} - \frac{b}{a}.$

25b. $x + \frac{1}{x} = \frac{a-b}{a+b} + \frac{a+b}{a-b}.$

26a. $\frac{a}{a+x} + \frac{a-x}{x} = \frac{11}{10}.$

26b. $\frac{a}{x-a} - \frac{x}{x+a} = \frac{7}{5}.$

27a. $\frac{x+a}{x-a} - \frac{x+b}{x-b} = 1.$

27b. $\frac{x+a}{x-b} - \frac{x-a}{x+b} = 1.$

28a. $\frac{a+4b}{x+2b} - \frac{a-4b}{x-2b} = \frac{4b}{a}.$

28b. $\frac{a+6b}{x+3b} + \frac{a-6b}{3b-x} = \frac{6b}{a}.$

29a. $\frac{1}{a} + \frac{1}{a+x} + \frac{1}{a+2x} = 0.$

29b. $\frac{a}{x-a} - \frac{2a}{x-2a} + \frac{3a}{x-3a} = 0.$

30. $\frac{x}{x+a} + \frac{2x}{x-a} = \frac{5a^2}{4(x^2-a^2)}.$

31. $\frac{1}{a+b+x} = \frac{1}{a} + \frac{1}{b} + \frac{1}{x}.$

32. $\frac{a(x-2)}{b} + \frac{a}{bx}\left(1 - \frac{b^2}{a^2}\right) = \frac{b(x-2)}{a}.$

33. $(a+b)(a-b)x^2 = ab(2ax - ab).$

34. $x^2 - \frac{cx}{a+b} - \frac{2c^2}{(a+b)^2} = 0.$

35. $\frac{2a+b-x}{2b+a-x} = \frac{a}{b} \cdot \frac{x+b}{x+a}.$

36. $\frac{4a+3b-x}{4b+3a-x} = \frac{2a+b}{2b+a} \cdot \frac{2a+3b+x}{2b+3a+x}.$

37. $\frac{x+a}{x-a} + \frac{x+b}{x-b} = \frac{a}{b} + \frac{b}{a}.$

38. $\frac{1}{x-a} + \frac{1}{x-b} + \frac{1}{x-c} = 0.$

39. $\frac{a+b-x}{a-b-x} = \frac{a-c+x}{a-c-x}.$

40. $\frac{(a-x)(a-b)+(x-b)^2}{(a-x)^2+(2x-a-b)(x-b)} = \frac{49}{19}.$

41. $\frac{a+c(a+x)}{a+c(a-x)} + \frac{a+x}{x} = \frac{a}{a-2cx}.$

42. $\frac{x+a}{x-a} + \frac{x+b}{x-b} + \frac{x+c}{x-c} = 3.$

43. $\frac{x^2+1}{a^n} = x(1 + a^{-2n}).$

44. $x^2 + a^n x = a^{3n}x + a^{4n}.$

45. $x^2\sqrt{a} = a\sqrt{2a} + (a - a\sqrt{2})x.$

Properties of the Roots of a Quadratic Equation

The roots of the reduced quadratic equation $x^2 + px + q = 0$ are real and distinct under the condition $p^2 > 4q$, real and equal under the condition $p^2 = 4q$, and imaginary under the condition $p^2 < 4q$. Similarly, the roots of the general equation $ax^2 + bx + c = 0$ are real and distinct under the condition $b^2 > 4ac$, real and equal under $b^2 = 4ac$, and imaginary under $b^2 < 4ac$.

Without solving, determine which of the following equations have real distinct, real equal, or imaginary roots:

46a. $x^2 + 6x + 5 = 0.$ **46b.** $x^2 - 6x + 8 = 0.$

47a. $x^2 - 10x + 25 = 0.$ **47b.** $x^2 - 14x + 49 = 0.$

48a. $x^2 + 4x + 5 = 0.$ **48b.** $x^2 - 9x + 20 = 0.$

49a. $x^2 + 8x + 25 = 0.$ **49b.** $x^2 + 11x + 130 = 0.$

50a. $x^2 + 2x - 120 = 0.$ **50b.** $x^2 + 3x - 180 = 0.$

51a. $x^2 + 24x + 144 = 0.$ **51b.** $x^2 + 30x + 225 = 0.$

52a. $12x^2 + 7x - 12 = 0.$ **52b.** $9x^2 - 12x + 4 = 0.$

53a. $4x^2 - 4x + 13 = 0.$ **53b.** $3x^2 + 12x + 13 = 0.$

54a. $25x^2 + 30x + 9 = 0.$ **54b.** $9x^2 - 42x + 49 = 0.$

55a. $2x^2 - 18x + 65 = 0.$ **55b.** $36x^2 + 48x + 61 = 0.$

In the reduced equation the sum of the roots equals the coefficient p taken with the opposite sign, and the product of the roots equals the coefficient q. In the general equation the sum of the roots equals the ratio of the coefficients $\frac{b}{a}$ taken with the opposite sign, and the product of the roots equals the ratio of the coefficients $\frac{c}{a}$. Using these remarks, one can determine the signs of the real roots.

Without solving the following equations, determine the signs of their roots, if the latter are real:

56a. $x^2 - 8x + 15 = 0.$ **56b.** $x^2 + 9x + 14 = 0.$

57a. $x^2 + 4x - 3 = 0.$ **57b.** $x^2 - 2x - 15 = 0.$

58a. $x^2 - 17x - 60 = 0.$ **58b.** $x^2 + x - 42 = 0.$

59a. $x^2 - 5x + 130 = 0.$ **59b.** $x^2 + 7x + 200 = 0.$

60a. $x^2 - 26x + 169 = 0.$ **60b.** $x^2 - 34x + 289 = 0.$

61a. $x^2 - 3x - 460 = 0.$ **61b.** $x^2 - 3x - 340 = 0.$

62a. $2x^2 + 5x + 2 = 0.$ **62b.** $3x^2 - 7x + 2 = 0.$

63a. $6x^2 - 5x - 6 = 0.$ **63b.** $9x^2 - 24x - 20 = 0.$

64a. $4x^2 + 2x + 1 = 0.$ **64b.** $9x^2 + 3x + 1 = 0.$

65a. $8x^2 + 4x - 1 = 0.$ **65b.** $26x^2 - 30x - 1 = 0.$

A quadratic trinomial of the form $x^2 + px + q$ always factors into the product $(x - x_1)(x - x_2)$, where x_1 and x_2 are the roots of the trinomial. A trinomial of the form $ax^2 + bx + c$ factors into the product $a(x - x_1)(x - x_2)$, differing from the previous by the factor a.

Factor the trinomials into products:

66a. $x^2 - 7x + 12.$ **66b.** $x^2 - 9x + 18.$

67a. $x^2 + 3x - 108.$ **67b.** $x^2 + 5x - 204.$

68a. $6x^2 + 5x - 6.$ **68b.** $15x^2 + 34x + 15.$

69a. $30x^2 + 37x + 10.$ **69b.** $21x^2 + 22x - 8.$

70a. $x^2 - 6x + 11.$ **70b.** $x^2 - 9x + 21.$

71a. $x^2 + 15x + 44.$ **71b.** $x^2 - 10x + 22.$

72a. $x^2 - ax - 6a^2.$ **72b.** $x^2 + ax - 2a^2.$

73a. $abx^2 - 2ax + a^2 - b^2$.

73b. $(a^2 + b^2)x^2 - 2b^2x + b^2 - a^2$.

74a. $x^2 - ax - a\sqrt{b} - b$.

74b. $x^2 + \sqrt{b}\,x - a^2 + a\sqrt{b}$.

75a. $abx^2 - 2a\sqrt{ab} \cdot x + a^2 - b^2$.

75b. $a^2b^2x^2 - 2ab^2\sqrt{b} \cdot x + b^3 - a^3$.

Using the connection between the coefficients and the roots of a quadratic equation, one can compose equations from their given roots. The equation is composed in reduced form. If the coefficients of the obtained equation turn out to be fractional, then, clearing the denominator, we obtain the equation in general form.

Compose quadratic equations from their given roots:

76a. 2 and 3.

76b. 7 and -5.

77a. -4 and 6.

77b. -8 and -5.

78a. -5 and 0.

78b. 8 and 0.

79a. 3 and -3.

79b. -7 and 7.

80a. $\frac{1}{2}$ and $-\frac{1}{4}$.

80b. $\frac{2}{3}$ and $\frac{1}{3}$.

81a. $-\frac{2}{3}$ and $-\frac{3}{2}$.

81b. $\frac{3}{7}$ and $\frac{7}{8}$.

82a. $\sqrt{6}$ and $-\sqrt{3}$.

82b. $\sqrt{2}$ and $-\sqrt{6}$.

83. $4 \pm \sqrt{3}$.

84. $-3 \pm \sqrt{-15}$.

85. $1 \pm \sqrt{-10}$.

86. $3a,\ -2b$.

87. $2a - b,\ a - 2b$.

88. $-\frac{a}{3},\ \frac{a}{2}$.

89. $a \pm b$.

90. $\frac{a}{b},\ \frac{b}{a}$.

91. $\frac{a-b}{a+b},\ 1$.

92. $\frac{b}{1-a},\ \frac{a}{1-b}$.

93. $a \pm \sqrt{b}$.

94. $\sqrt{a} \pm \sqrt{-b}$.

95a. Compose the equation whose roots would be the reciprocals of the roots of the equation $x^2 + px + q = 0$.

95b. The same for an equation of the form $ax^2 + bx + c = 0$.

96a. Compose the equation whose roots would be m times the roots of the equation $ax^2 + bx + c = 0$.

96b. The same for an equation of the form $x^2 + px + q = 0$.

97. Compose the equation whose roots would be $\frac{p}{2}$ greater than the roots of the equation $x^2 + px + q = 0$.

98. Compose the equation whose roots would be the sum and the product of the roots of the equation $ax^2 + bx + c = 0$.

99. Express the sum of the cubes of the roots of the equation $ax^2 + bx + c = 0$.

100. Express the incomplete square of the sum of the roots of the equation $x^2 + px + q = 0$.

101. For what value of b does the equation $4x^2 + bx + 25 = 0$ have equal roots?

102a. Show that if the discriminant of the quadratic equation $ax^2 \pm bx + c = 0$ equals zero, then the left side of this equation is a perfect square.

102b. The same for an equation of the form $x^2 \pm px + q = 0$.

103. For what positive values of c are the roots of the equation $3x^2 - 18x + c = 0$ real, and for what imaginary?

104. Solve the equation $ax^2 + bx = 0$ by the formula for the complete quadratic equation.

105. Solve the equation $ax^2 + c = 0$ by the formula for the complete quadratic equation.

106. In the equation $x^2 - 6x + q = 0$ determine the value of q for which the roots α and β satisfy the relation $3\alpha + 2\beta = 20$.

107. In the equation $x^2 - 5x + q = 0$ determine the value of q for which the roots α and β satisfy the relation $3\alpha + 5\beta = 17$.

108. What relation must exist in the equation $x^2 + px + q = 0$ between the coefficients p and q so that one root of this equation is m times the other?

109. In the formula for solving the equation $x^2+px+q = 0$, remove the irrationality in the numerator of the fraction.

Composing Literal Quadratic Equations

110. Find two numbers whose product is p and quotient q.

111. On dividing one number by another, the quotient is a and the remainder b; the product of these two numbers equals c. Find these numbers.

112. Decompose the number a into two factors whose difference equals b.

113. Find two numbers the sum of whose squares is S and whose ratio is $p : q$.

114. The sum of a proper fraction with its reciprocal equals a. Find the value of the fraction; determine for what values of a the problem has a solution.

115. One part of a sum of a rubles brings annually b rubles, and the other c rubles of profit. What percent of profit does each part give, if from the second one obtains one percent more than from the first?

116. A segment of length a is divided into two parts, one of which is the mean proportional between the whole segment and the second part. Determine the lengths of the parts of the segment.

117. Divide a rectilinear segment of length a into two parts such that twice the area of the square constructed on one part equals the area of the rectangle whose dimensions are the other part and the given segment itself.

118. The perimeter of a rectangle is $2p$, the area is S. Determine the sides of the rectangle; investigate the solution and find out for what relation between p and S the rectangle becomes a square.

119. In an isosceles triangle the lateral side equals n, and the altitude is less than the base by m. Find the base and the altitude of the isosceles triangle.

120. In a right triangle, from the vertex of the right angle a perpendicular is dropped to the hypotenuse. Determine the parts into which this perpendicular divides the hypotenuse, if the length of the perpendicular is h and of the hypotenuse c.

121. If the radius of a circle is increased by a centimeters, the area of the circle increases n times. Determine the radius of the circle.

122. Along the sides of a right angle, from its vertex, two bodies begin to move at the same time. The speed of one is V_1 centimeters per second, the speed of the other V_2 centimeters per second. After what time will the distance between these bodies be d centimeters?

123. The resultant of two forces applied to a point at an angle of $120°$ equals P; the ratio of these forces is $m : n$. Determine the component forces.

124. An electric lamp of m candles hung over a table. When it burned out, it was replaced by a lamp of n candles and, so that the degree of illumination would not change, lowered by h centimeters. At what height over the table does the lamp now hang?

125. Two magnets A and B are at a distance of d centimeters from each other. Magnet B is n times stronger than magnet A. At what distance from magnet A is the point at which the attraction of both magnets is equal, given that the force of attraction of a magnet is inversely proportional to the square of its distance from the attracted body?

126. Two water pipes with diameters a and b are to be replaced by one pipe with the same carrying capacity. What must be the diameter of this pipe?

127. A stone falls into a well, and the observer hears the sound of its fall n seconds after the start of the fall. Find the depth of the well, if the speed of sound equals 330 m per second and $g = 9.8$ m per second squared.

128. From a rectangular piece of sheet metal with dimensions a and b, it is required to make an open box so that: 1) the area of the walls equals the area of the bottom, 2) the area of the bottom has a given size S. Determine the conditions for the validity of the roots.

129. A railway track has a length of a kilometers. If the speed of the train is increased by b kilometers per hour, the train will spend c hours less on covering the track than now. How much time does the train spend on covering the whole track, and what is its speed?

130. Two automobiles set out simultaneously from place A to place B. One of them travels m kilometers more per hour than the other, and therefore arrives at place B n hours earlier than the other. The distance between A and B equals

P kilometers. How many kilometers per hour does each automobile travel?

131. Two rolling mills can, working simultaneously, roll a kilograms of iron in T hours. If only the first of the mills works, then rolling the stated quantity of iron will require t hours more time than the work of the second mill alone. In what time can each of the mills roll a kilograms of iron, working separately?

The Graph of a Quadratic Function. Graphical Solution of a Quadratic Equation

The graph of the equation $y = ax^2$ is a parabola tangent to the X-axis at the point O and symmetrically situated with respect to the Y-axis. If $a > 0$, the parabola $y = ax^2$ lies on the side of the X-axis where the positive values of the ordinates lie; if $a < 0$, conversely, the parabola lies on the side where the negative values lie. The graph of the equation $y = ax^2 + bx + c$ is a parabola whose axis is parallel to the Y-axis and whose branches are directed in the positive or negative direction of the Y-axis, depending on whether $a > 0$ or $a < 0$.

A quadratic equation can be solved graphically in two ways. *First method.* Having constructed the parabola $y = ax^2 + bx + c$ by points, we determine the roots of the equation $ax^2 + bx + c = 0$ as the abscissas of the points of intersection of the parabola with the X-axis. If the parabola intersects the X-axis in two points, the equation has two distinct real roots; if it touches the X-axis, two equal real roots; if it does not intersect the X-axis, two imaginary (conjugate) roots (Fig. 1). *Second method.* Rewrite the equation $ax^2 + bx + c = 0$ in the form $ax^2 = -bx - c$. Setting $ax^2 = y$, we obtain the system $y = ax^2$; $y = -bx - c$. Since the roots of these equations must be the same, they must be the coordinates of points lying simultaneously on both graphs, i.e. the coordinates of the points of intersection of these graphs. Constructing on one drawing the graph of $y = ax^2$ and the graph of $y = -bx - c$, we find the coordinates of their points of intersection (Fig. 2). If the parabola $y = ax^2$ has two common points with the line $y = -bx - c$, the equation $ax^2 + bx + c = 0$ has two distinct real roots; if one common point, two equal real roots; if no common points, two imaginary (conjugate) roots (Fig. 3).

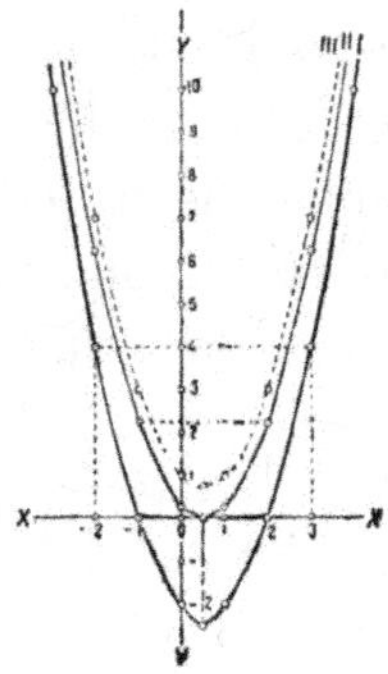

I. $x^2 - x - 2 = 0.$
II. $x^2 - x + \frac{1}{4} = 0.$
III. $x^2 - x + 1 = 0.$

Fig. 1.

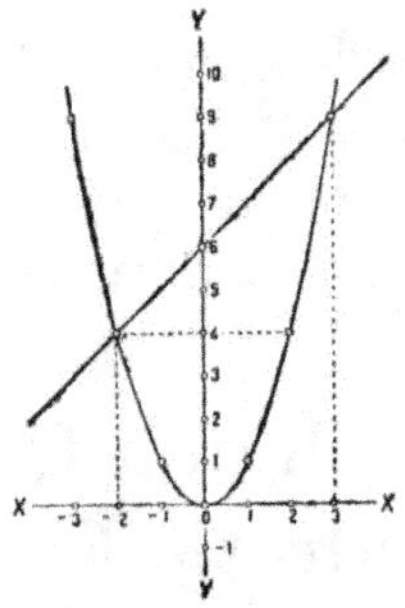

$x^2 - x - 6 = 0.$
$y = x^2.$
$y = x + 6.$

Fig. 2.

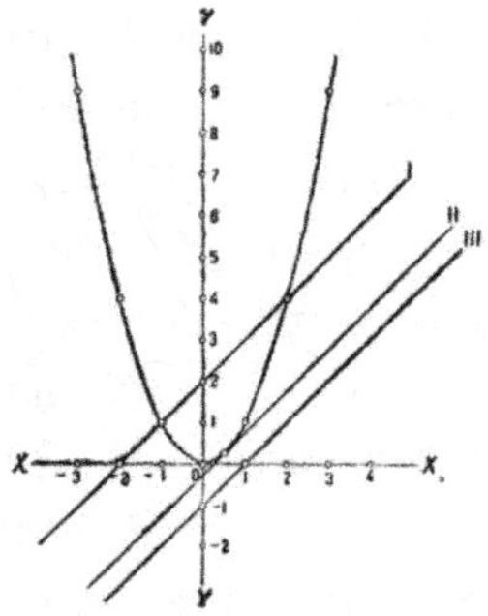

I. $y = x^2; \quad y = x + 2.$
II. $y = x^2; \quad y = x + \frac{1}{4}.$
III. $y = x^2; \quad y = x - 1.$

Fig. 3.

132. Construct the graphs of the functions: $y = x^2$; $y = \frac{1}{2}x^2$; $y = 2x^2$.

133. Construct the graphs of the functions: $y = -x^2$; $y = -\frac{1}{4}x^2$; $y = -3x^2$.

134. Construct the graphs of the functions: $y = x^2 + 1$; $y = x^2 - 2$.

135. Construct the graphs of the functions: $y = x^2 + 3x$; $y = x^2 - 4x$.

136. Construct the graph of the function: $y = x^2 - 3x + 2$.

137. Construct the graph of the function: $y = x^2 + x - 12$.

138. Construct the graph of the function: $y = x^2 - 4x + 4$.

139. Construct the graph of the function: $y = 2x^2 - 2x + 3$.

140. Construct the graph of the dependence of the area of a square on its side.

141. Construct the graph of the dependence of the area of a circle on its radius.

142. Solve graphically the equation $x^2 - 3x = 0$.

143. Solve graphically the equation $x^2 + 4x = 0$.

144. Solve graphically the equation $x^2 - 4 = 0$.

145. Solve graphically the equation $x^2 - 2x - 3 = 0$.

146. Solve graphically the equation $x^2 - 5x + 6 = 0$.

147. The volume V of one gram of water at temperature t°C is approximately expressed by the formula $V = 1 + 8.38 \cdot 10^{-6}(t - 4)^2$. Construct the graph of the dependence of V on t and determine for what value of t the volume V attains its smallest value.

148. A shell from an anti-aircraft gun flies out of the muzzle vertically upward with an initial speed of 125 m/sec; the path S (in meters) it has travelled, if air resistance is neglected, is expressed by the formula $S = 125t - 4.905t^2$, where t is the time elapsed from the start of motion. Construct the graph of the dependence of S on t; determine the greatest height of ascent of the shell and at what moment it will be reached.

Chapter XII

Equations of Higher Degrees

An equation of higher degree is any equation of degree higher than the second. Every algebraic equation of degree n has n roots, which need not all be distinct, and the left-hand side of the equation factors into n linear factors (Gauss's theorem).

The general form of an equation of the 3rd degree is $ax^3 + bx^2 + cx + d = 0$. If we divide both sides by a, we obtain the reduced equation, written in the form $x^3 + px^2 + qx + r = 0$.

The general form of an equation of the 4th degree is

$$ax^4 + bx^3 + cx^2 + dx + e = 0,$$

and the reduced equation of the 4th degree is

$$x^4 + px^3 + qx^2 + rx + s = 0.$$

The Biquadratic Equation

A biquadratic equation is an equation of the 4th degree in which the terms with odd powers of the unknown are absent, i.e. an equation of the form

$$ax^4 + bx^2 + c = 0.$$

This equation can be regarded as quadratic, not in x but in x^2; x^2 can be found by the formula for solving a quadratic equation, and by extracting the square root of the obtained result we determine x. Thus the formula for solving the biquadratic equation is

$$x = \pm\sqrt{\frac{-b \pm \sqrt{b^2 - 4ac}}{2a}}.$$

This formula gives four roots of the biquadratic equation, namely:

$$x_1 = +\sqrt{\frac{-b+\sqrt{b^2-4ac}}{2a}}; \quad x_2 = -\sqrt{\frac{-b+\sqrt{b^2-4ac}}{2a}};$$

$$x_3 = +\sqrt{\frac{-b-\sqrt{b^2-4ac}}{2a}}; \quad x_4 = -\sqrt{\frac{-b-\sqrt{b^2-4ac}}{2a}}.$$

1a. $x^4 - 5x^2 + 4 = 0.$

1b. $x^4 + 12x^2 - 64 = 0.$

2a. $x^4 + 12x^2 + 32 = 0.$

2b. $x^4 + 9x^2 + 20 = 0.$

3a. $5x^4 + x^2 - 4 = 0.$

3b. $3x^4 - x^2 - 2 = 0.$

4a. $12x^4 + x^2 - 6 = 0.$

4b. $6x^4 - x^2 - 15 = 0.$

5. $x^4 - (a-b)x^2 - ab = 0.$

6. $x^2 - \frac{4a^2}{x^2} = 4 - a^2.$

7. $a^2b^2x^4 - (a^4 + b^4)x^2 + a^2b^2 = 0.$

8a. What is the sum of the roots of a biquadratic equation?

8b. What is the product of the roots of a biquadratic equation?

9. Factor the trinomial $4x^4 - 17x^2 + 4$.

10. Form an equation whose roots are ± 1 and ± 3.

The Binomial Equation

A reduced binomial equation is an equation of the form $x^n \pm a = 0$. To solve such equations one sets $x = \sqrt[n]{a} \cdot z$, whereby the given equations reduce to the simpler $z^n - 1 = 0$ and $z^n + 1 = 0$. For several small values of n these are solved by factoring their left-hand sides, after which the found values of z are multiplied by $\sqrt[n]{a}$. Equations of the general form $ax^n \pm b = 0$ are transformed into reduced ones by dividing by the coefficient a and are solved in the same way.

11a. $x^3 = -1.$

11b. $x^3 = 1.$

12a. $x^3 = 8.$

12b. $x^3 = -8.$

13a. $x^3 - a^3 = 0.$

13b. $x^3 + a^3 = 0.$

14a. $x^4 - 16 = 0.$

14b. $x^4 - 81 = 0.$

15a. $x^4 + 81 = 0.$

15b. $x^4 + 16 = 0.$

16a. $x^6 - 64 = 0.$

16b. $x^6 - 729 = 0.$

17a. $125x^3 + 8 = 0.$

17b. $125x^3 - 27 = 0.$

18a. $81x^4 + 4 = 0.$

18b. $16x^4 - 25 = 0.$

19a. $a^8x^8 - b^8 = 0.$

19b. $b^8x^8 - a^8 = 0.$

The Trinomial Equation

A trinomial equation is an equation of the form

$$ax^{2n} + bx^n + c = 0; \quad n \geq 2.$$

Its solution reduces to solving two binomial equations by the substitution $x^n = z$, which turns the given equation into a quadratic and allows two values of z to be found.

20a. $x^6 - 3x^3 + 2 = 0.$

20b. $x^6 + 4x^3 + 3 = 0.$

21a. $x^6 - 28x^3 + 27 = 0.$

21b. $x^6 - 19x^3 - 216 = 0.$

22. $4 = -\dfrac{x^{-6}+3}{x^{-3}}.$

23. $\dfrac{8-x^{-2}}{x^{-4}} = \dfrac{x^{-2}-(3x+5x^2)}{5}.$

24. $x^{\frac{6}{5}} + 8 = 9\sqrt[5]{x^3}.$

25. $x^{\frac{6}{5}} - 7 = 6\sqrt[5]{x^3}.$

26. $\dfrac{a^3}{9x^{-3}+7a^3} - \dfrac{3x^{-3}}{a^3-5x^{-3}} = 0.$

27. $x^{10} - 10x^5 + 9 = 0.$

Equations Whose Left-Hand Side Factors

Sometimes the left-hand side of an equation of higher degree can be factored — into linear or nonlinear factors. Then it is appropriate to apply the rule that a product of several factors can equal zero only when one of them equals zero. Setting each factor equal to zero, we split the given equation into a series of equations whose solution yields the roots of the original equation.

28a. $x(2x-7) = 0.$

28b. $x(ax-b) = 0.$

29a. $(5x-8)(4x+7)(x+9) = 0.$

29b. $(ax+b)(cx-d)(x-e) = 0.$

30a. $x^3 - 3x = 2.$

30b. $x^3 + 4 = 3x^2.$

31a. $x^3 + 6 = 7x.$

31b. $x^3 + 12 = 13x.$

32a. $x^3 + x^2 = x + 1.$

32b. $x^3 - x^2 = x - 1.$

33a. $x^3 - 5x^2 = x - 5.$

33b. $x^3 + 2x^2 = 4x + 8.$

34a. $x^3 + 2x^2 - 2x + 3 = 0.$

34b. $x^3 - 4x^2 - 4x - 5 = 0.$

35. $x^3 + 8x^2 + 15x + 18 = 0.$

36. $x^4 + x^3 = -2x + 4.$

37. $x^4 + 2x^3 - 13x^2 - 14x + 24 = 0.$

38. $x^4 - 2x^3 - 8x^2 + 19x - 6 = 0.$

39. $nx^3 + x + n + 1 = 0.$

40. $(a+x)^3 = (3a-x)^3.$

41. $x^3 + (b^2-a^2)x + ab^2 = 0.$

42. $\dfrac{a^2+ax+x^2}{a^2-ax+x^2} = \dfrac{a^2}{x^2}.$

The Reciprocal, or Symmetric, Equation

A reciprocal (symmetric) equation is an equation of any degree in which the coefficients of terms equidistant from the beginning and the end are equal. Only equations of the 3rd, 4th, and 5th degree are solved fully in the most general form. A cubic equation is solved by simply factoring its left-hand side. The left-hand side of a

5th-degree equation also splits into two equations, one linear and the other of the 4th degree, again reciprocal. A reciprocal equation of the 4th degree of the first kind is

$$ax^4 + bx^3 + cx^2 + bx + a = 0.$$

Dividing both sides of this equation by x^2, we obtain

$$ax^2 + bx + c + \frac{b}{x} + \frac{a}{x^2} = 0.$$

Grouping the terms with equal coefficients, we get

$$a\left(x^2 + \frac{1}{x^2}\right) + b\left(x + \frac{1}{x}\right) + c = 0.$$

Replacing $x + \frac{1}{x}$ by a new unknown y, we obtain

$$x^2 + \frac{1}{x^2} = y^2 - 2$$

and thereby reduce the reciprocal equation to a quadratic in y.

A reciprocal equation of the 4th degree of the second kind is

$$ax^4 + bx^3 + cx^2 - bx + a = 0.$$

It is solved in the same way as the equation of the first kind, with the sole difference that y will replace $x - \frac{1}{x}$, and the expression $x^2 + \frac{1}{x^2}$ will be expressed through $y^2 + 2$.

A reciprocal equation of the 5th degree, as already stated, reduces — after separating the linear factor — to a reciprocal equation of the 4th degree. Every reciprocal equation of even degree reduces to an equation of half the degree by a method analogous to the solution of the 4th-degree equation. But this new equation of half the degree is, generally speaking, not reciprocal. For example:

$$ax^6 + bx^5 + cx^4 + dx^3 + cx^2 + bx + a = 0;$$

$$ax^3 + bx^2 + cx + d + \frac{c}{x} + \frac{b}{x^2} + \frac{a}{x^3} = 0;$$

$$a\left(x^3 + \frac{1}{x^3}\right) + b\left(x^2 + \frac{1}{x^2}\right) + c\left(x + \frac{1}{x}\right) + d = 0;$$

$$x + \frac{1}{x} = y; \quad x^2 + \frac{1}{x^2} = y^2 - 2; \quad x^3 + \frac{1}{x^3} = y^3 - 3y;$$

$$a(y^3 - 3y) + b(y^2 - 2) + cy + d = 0$$

and finally

$$ay^3 + by^2 - (3a - c)y - (2b - d) = 0.$$

If this equation is solvable, then the reciprocal equation of the 6th degree is also solvable. Finally, incomplete reciprocal equations are possible; they are often solvable by factoring the left-hand side.

43a. $x^3 - x^2 + x - 1 = 0.$

43b. $ax^3 + bx^2 + bx + a = 0.$

44a. $2x^4 - 5x^3 + 5x - 2 = 0.$

44b. $3x^4 - 10x^3 + 10x - 3 = 0.$

45a. $6x^4 - 5x^3 - 5x - 6 = 0.$

45b. $12x^4 + 7x^3 + 7x - 12 = 0.$

46a. $ax^4 - bx^3 + bx - a = 0.$

46b. $ax^4 - bx^3 - bx + a = 0.$

47a.
$6x^4 + 5x^3 - 38x^2 + 5x + 6 = 0.$

47b.
$6x^4 - 35x^3 + 62x^2 - 35x + 6 = 0.$

48a. $2x^4 + x^3 - 11x^2 + x + 2 = 0.$

48b. $2x^4 - 3x^3 - x^2 - 3x + 2 = 0.$

49a. $4x^4 - 33x^3 + 33x + 4 = 0.$

49b. $6x^4 + 73x^3 - 73x + 6 = 0.$

50. $6x^4 + 7x^3 - 36x^2 - 7x + 6 = 0.$

51. $2x^5 + 5x^4 - 13x^3 - 13x^2 + 5x + 2 = 0.$

52. $15x^5 + 34x^4 + 15x^3 - 15x^2 - 34x - 15 = 0.$

53. $x^6 - 10x^5 + 27x^4 - 20x^3 + 27x^2 - 10x + 1 = 0.$

54. $2x^6 - x^5 - 8x^4 + 8x^2 + x - 2 = 0.$

55a. Prove that the roots of a reciprocal equation of the 4th degree of the first kind are pairwise reciprocal in magnitude.

55b. Prove that the roots of a reciprocal equation of the 4th degree of the second kind are pairwise reciprocal both in magnitude and in sign.

56a. Factor the symmetric polynomial
$2x^4 - 9x^3 + 14x^2 - 9x + 2.$

56b. Factor the symmetric polynomial
$30x^4 - 17x^3 - 228x^2 + 17x + 30.$

57a. Form an equation with roots 2; $\frac{1}{2}$; 3 and $\frac{1}{3}$.

57b. Form an equation with roots 2; $-\frac{1}{2}$; 3 and $-\frac{1}{3}$.

Chapter XIII

Irrational Equations

An *irrational equation* is an equation in which the unknown appears under a radical sign. To solve such an equation, it must be replaced by another that contains no roots of expressions involving the unknown. This is achieved by raising to a power, applied once or several times in succession.

Since raising to a power introduces extraneous solutions, after solving an irrational equation one must check each root by substituting it into the equation that was originally raised to a power. If the tested root does not satisfy the equation being checked, then it is not a root of the given equation and must belong to one of the additional equations, of which there are always as many as the number of times raising to a power was carried out during the solution.

1. $5+\sqrt{6-x}=7.$ $\qquad$ $x+\sqrt{16+x^2}=8.$

2. $\sqrt{5+\sqrt{x-4}}=3.$ $\qquad$ $\sqrt{17-\sqrt{x-8}}=4.$

3. $\sqrt{x+1}+\sqrt{2x+3}=1.$ $\qquad$ $\sqrt{2x-1}+\sqrt{x-1}=1.$

4. $\sqrt{3x+4}+\sqrt{x+2}=8.$ $\qquad$ $\sqrt{x+3}+\sqrt{3x-3}=10.$

5. $\sqrt{22-x}-\sqrt{10-x}=2.$ $\qquad$ $\sqrt{x+20}-\sqrt{x-1}=3.$

6. $2\sqrt{x+18}+\sqrt{4x-3}=15.$ $\qquad$ $\sqrt{x-7}-\sqrt{x+1}=-2.$

7. $\sqrt{2x+1}+\sqrt{x-3}=2\sqrt{x}.$ $\qquad$ $\sqrt{x+3}+\sqrt{x+8}=5\sqrt{x}.$

8. $\sqrt{3x-3}+\sqrt{5x-19}=\sqrt{3x+4}.$ $\qquad$ $\sqrt{2x+1}+\sqrt{7x-27}=\sqrt{3x+4}.$

9. $\sqrt{1+x\sqrt{x^2+12}}=1+x.$ $\qquad$ $\sqrt{1+x\sqrt{x^2-24}}=x-1.$

10. $x=-2+\sqrt{4+x\sqrt{36+x^2}}.$ $\qquad$ $x=1-\sqrt{1-x\sqrt{16+x^2}}.$

11. $\frac{2}{x}+2=\sqrt{4+\frac{1}{x}\sqrt{64+\frac{144}{x^2}}}.$

12. $1-\frac{1}{x}=\sqrt{1-\frac{1}{x}\sqrt{4-\frac{7}{x^2}}}.$

13. $\frac{5}{x+\sqrt{5+x^2}}-\frac{5}{x-\sqrt{5+x^2}}=6.$

14. $\frac{4}{x+\sqrt{4-x^2}}+\frac{4}{x-\sqrt{4-x^2}}=\frac{12}{7}.$

15. $\frac{x-1}{1+\sqrt{x}}=4-\frac{1-\sqrt{x}}{2}.$

16. $\sqrt{5x} - \frac{4}{\sqrt{3x+1}} = \sqrt{3x+1}.$

17. $\frac{\sqrt{2x^2+1}+\sqrt{x-1}}{\sqrt{2x^2+1}-\sqrt{x-1}} = 2.$

18. $\frac{\sqrt{3x^2+1}-\sqrt{2x+1}}{\sqrt{3x^2+1}+\sqrt{2x+1}} = \frac{2}{5}.$

19. $\sqrt{x+\sqrt{x}} - \sqrt{x-\sqrt{x}} = \frac{3}{2}\sqrt{\frac{x}{x+\sqrt{x}}}.$

20. $\frac{x+1-\sqrt{2x+1}}{x+1+\sqrt{2x+1}} = \frac{\sqrt{2x+1}+1}{\sqrt{2x+1}-1}.$

21. $(x+2)^{\frac{1}{2}} + x^{\frac{1}{2}} = 4(x+2)^{-\frac{1}{2}}.$

22. $[5+(x-4)^{\frac{1}{2}}]^{\frac{1}{2}} = 3.$

23. $\sqrt[3]{x} + 2\sqrt[3]{x^2} - 3 = 0.$

24. $\sqrt{x-3} + 6 = 5\sqrt[4]{x-3}.$

25. $x^2 + \sqrt{x^2-9} = 21.$

26. $3x^2 + 15x + 2\sqrt{x^2+5x+1} = 2.$

27. $x + \sqrt{2ax+x^2} = a.$ $2a - \sqrt{2ax+x^2} = x.$

28. $\sqrt{x} + \sqrt{a-x} = \sqrt{a}.$ $\sqrt{a+x} + \sqrt{a-x} = \sqrt{2a}.$

29. $\sqrt{3x+a+2b} - \sqrt{3x+a-2b} = 2\sqrt{x-a}.$ $\sqrt{5x-3a+4b} + \sqrt{5x-3a-4b} = 2\sqrt{x+a}.$

30. $\sqrt{a-bx} + \sqrt{c-dx} = \sqrt{a+c-(b+d)x}.$

31. $\sqrt{a+x} + \sqrt{2a+x} = \frac{a}{\sqrt{a+x}}.$

32. $\sqrt{a+\sqrt{x}} - \sqrt{a-\sqrt{x}} = \sqrt{a}.$

33. $\frac{1}{a} - \frac{1}{x} = \sqrt{\frac{1}{a^2} - \sqrt{\frac{4}{a^2x^2} - \frac{7}{x^4}}}.$

34. $\frac{\sqrt{a+x}+\sqrt{a-x}}{\sqrt{a+x}-\sqrt{a-x}} = \frac{a}{x}.$

35. $\frac{\sqrt{ax+b}+\sqrt{ax}}{1+\sqrt{ax-b}} = \frac{\sqrt{ax+b}-\sqrt{ax}}{1-\sqrt{ax-b}}.$

36. $\frac{\sqrt{a+x}+\sqrt{x-a}}{\sqrt{b+2x}+\sqrt{b-x}} = \frac{\sqrt{a+x}-\sqrt{x-a}}{\sqrt{b+2x}-\sqrt{b-x}}.$

37. $\frac{\sqrt{a-x}+\sqrt{x-b}}{\sqrt{a-x}} = \frac{\sqrt{a-x}-\sqrt{x-b}}{\sqrt{x-b}}.$

Chapter XIV

Systems of Equations of Degree Higher than the First

Solving Systems of Equations

To solve a system of equations $Ax^2+Bxy+Cy^2+Dx+Ey+F=0$ and $ax+by=c$, of which one is of the second and the other of the first degree, we express y through x from the second equation and substitute the obtained expression $y=\frac{c-ax}{b}$ into the first. A quadratic equation of the form $Mx^2+Nx+P=0$ results. Solving the latter, we find two values x_1 and x_2 and, substituting them into the expression for y, obtain the corresponding values y_1 and y_2. As a result, two systems of solutions are obtained.

To solve two equations of the second degree $Ax^2+Bxy+Cy^2+Dx+Ey+F=0$ and $A_1x^2+B_1xy+C_1y^2+D_1x+E_1y+F_1=0$, we first eliminate from them the square of one unknown, for example y. For this we multiply the first equation by C_1, the second by C, and subtract one from the other. We obtain an auxiliary equation, which for brevity we represent in the form $ax^2+bxy+dx+ey+f=0$. Using the fact that the obtained equation contains only the first power of y, we express y through x from it in rational form: $y=-\frac{ax^2+dx+f}{bx+e}$. The obtained expression for y is inserted into one of the given equations. Then an equation in x alone of the fourth degree is composed. If the latter is solved, four values of x will be found, and inserting each of them into the preceding expression for y through x, we obtain four corresponding values of y. In the case where the square of one of the unknowns does not enter one of the equations, the computation is simplified.

Solve the systems of equations:

1a. $\begin{cases} x^2-y^2=32; \\ x-2y=2. \end{cases}$ **1b.** $\begin{cases} x^2+y^2=41; \\ y-x=1. \end{cases}$

2a. $\begin{cases} 2x^2-2xy+x=-9; \\ 2y-3x=1. \end{cases}$ **2b.** $\begin{cases} x^2+3xy-y^2=92; \\ x+3y=18. \end{cases}$

3a. $\begin{cases} x^2+6xy+8y^2=91; \\ x+3y-10=0. \end{cases}$

3b. $\begin{cases} 2x^2+10xy+17y^2=218; \\ 2x+5y-20=0. \end{cases}$

4a. $\begin{cases} x^2+2xy-4y^2-5x+4=0; \\ x-y=2. \end{cases}$

4b.
$\begin{cases} 2x^3-xy+3y^2-7x-12y+1=0; \\ x+1=y. \end{cases}$

5a. $\begin{cases} xy=0.5; \\ \frac{x}{y}=2. \end{cases}$

5b. $\begin{cases} \frac{1}{xy}=5; \\ \frac{x}{y}=20. \end{cases}$

6a. $\begin{cases} xy=99; \\ \frac{x+y}{x-y}=10. \end{cases}$

6b. $\begin{cases} \frac{x+y}{y}=2; \\ 1+\frac{xy}{3}=4. \end{cases}$

7a. $\begin{cases} x-6=y+6; \\ \frac{3}{5}x-y=y(y-1). \end{cases}$

7b. $\begin{cases} x-y=4; \\ \frac{x}{3}=\frac{7}{y}. \end{cases}$

8a. $\begin{cases} x+y=120; \\ x^2+y^2=10400. \end{cases}$

8b. $\begin{cases} \frac{x-y}{2}=1; \\ x^2-y^2=0. \end{cases}$

9a. $\begin{cases} x^2+y^2=58.4; \\ \frac{11x-3y}{x-1}=10. \end{cases}$

9b. $\begin{cases} x+y=2; \\ \frac{x^2-2+y^2}{2}=1. \end{cases}$

10a. $\begin{cases} (x-2)(y-3)=1; \\ \frac{x-2}{y-3}=1. \end{cases}$

10b. $\begin{cases} \frac{(x+y+1)(x+y-1)}{xy+57}=3; \\ \frac{1}{2x}-\frac{1}{2y}=-\frac{1}{xy}. \end{cases}$

11a. $\begin{cases} x-y=a; \\ \frac{89a^2-(5y-x)y}{(x+y)^2}=3. \end{cases}$

11b. $\begin{cases} 2x+y=5a; \\ \frac{x(x-y)+3y^2}{a(a-1)+3}=5. \end{cases}$

12a. $\begin{cases} x-y=\frac{2a}{a^2-1}; \\ \frac{1}{y}-\frac{1}{x}=\frac{2}{a}. \end{cases}$

12b. $\begin{cases} (b+y)x=b^2; \\ \frac{a^2b-b}{a^2b-y}=\frac{b}{b-x}. \end{cases}$

13a. $\begin{cases} \frac{x^2}{a^2}+\frac{y^2}{b^2}=1; \\ \frac{x}{a}+\frac{y}{b}=1. \end{cases}$

13b. $\begin{cases} \frac{a}{x^2}-\frac{b}{y^2}=0; \\ \frac{1}{x}+\frac{1}{y}=m. \end{cases}$

14a. $\begin{cases} x+y=a; \\ x^2+y^2=bxy. \end{cases}$

14b. $\begin{cases} \frac{x+y}{x-y}=\frac{a}{b}; \\ xy=c^2. \end{cases}$

15a. $\begin{cases} xy=a; \\ \frac{x}{y}=b. \end{cases}$

15b. $\begin{cases} \frac{x^2}{a^2}+\frac{y^2}{b^2}=2; \\ \frac{bx+ay}{bx-ay}=\frac{m}{n}. \end{cases}$

16a. $\begin{cases} x^2+3xy=18; \\ xy+4y^2=7. \end{cases}$

16b. $\begin{cases} x^2-xy+y^2=21; \\ 2xy-y^2=15. \end{cases}$

17a. $\begin{cases} x+y-x^2=0; \\ 3y-x-y^2=0. \end{cases}$

17b. $\begin{cases} 4x-4y-xy=0; \\ 2x^2+2y^2-5xy=0. \end{cases}$

18a.
$\begin{cases} 6x^2+xy-y^2-3x-4y=15; \\ 4xy-y^2-3x^2+15x-7y=18. \end{cases}$

18b.
$\begin{cases} 6x+21y-2x^2-27xy-6y^2=4; \\ 9xy+3y^2-2x^2+6x-6y=4. \end{cases}$

19a. $\begin{cases} 3x^2+2xy+y^2=43; \\ x^2+2xy+3y^2=33. \end{cases}$

19b. $\begin{cases} 3x^2-xy+4y^2=14; \\ 2x^2-xy+2y^2=8. \end{cases}$

20a. $\begin{cases} x^2 + xy + 2y^2 = 74; \\ 2x^2 + 2xy + y^2 = 73. \end{cases}$

20b. $\begin{cases} 3x^2 - 4xy + 2y^2 = 17; \\ y^2 - x^2 = 16. \end{cases}$

21a. $\begin{cases} \frac{1}{5}x^2 + \frac{1}{2}y^2 = 19.7; \\ \frac{(x+2)(x-2)}{(y+7)(7-y)} = \frac{4}{3}. \end{cases}$

21b. $\begin{cases} \frac{(x+1)(x-1)-78.3}{78.3-(y+1)(y-1)} = 1; \\ x + \frac{10.68}{y-x} + y = 0. \end{cases}$

22a. $\begin{cases} 3x^2 - 4x + y^2 = 40; \\ 2x^2 + 3x + y^2 = 52. \end{cases}$

22b. $\begin{cases} xy + x^2 = 18; \\ 2xy - x^2 = x + 6. \end{cases}$

23. $\begin{cases} \frac{2x+3xy-6y}{8} = 1; \\ \frac{x-2xy-3y}{3} = -1. \end{cases}$

24a. $\begin{cases} 8(x+y) - 7(xy+1) = 0; \\ 4(x-y) - (xy-1) = 0. \end{cases}$

24b. $\begin{cases} 6(x^2-1) - y(3x-y) = 13; \\ 4(x^2-1) - y(2x-y) = 9. \end{cases}$

25a. $\begin{cases} x^2 + y^2 + x + y = 68; \\ x^2 - y^2 + x - y = 44. \end{cases}$

25b. $\begin{cases} x^2 - 4y^2 - xy + 5y = 1; \\ x^2 + 3y^2 - xy - 4y = -1. \end{cases}$

26a. $\begin{cases} x^2 + xy = a; \\ y^2 + xy = b. \end{cases}$

26b. $\begin{cases} x^2 + xy + y^2 = 2a; \\ x^2 - xy + y^2 = 2b. \end{cases}$

27a. $\begin{cases} x(a-x) = y(a-y); \\ xy = a^2. \end{cases}$

27b. $\begin{cases} x - \frac{1}{y} = a; \\ y - \frac{1}{x} = \frac{1}{a}. \end{cases}$

28a. $\begin{cases} x^2 - xy = 2(a+1); \\ y^2 - xy = 2(1-a). \end{cases}$

28b. $\begin{cases} a^2x^2 + y^2 = 2a^4; \\ xy - a^3 = 0. \end{cases}$

29a. $\begin{cases} (x-a)(y-b) = 0; \\ \frac{x(x-b)-y(x-y)}{a^2+b^2} = 1. \end{cases}$

29b. $\begin{cases} (x+y+a)(x-y+a) = 0; \\ x^2 - \frac{1}{2}(a^4 + y^2 + 1) - x = a - y. \end{cases}$

30a. $\begin{cases} x^2 - (a+x)y - a^2 = 0; \\ \frac{x^2-2(x-y)+y^2}{5a-2} = a. \end{cases}$

30b. $\begin{cases} x^2 + y^2 + x + y = a^2(a^2+1); \\ a^2 + xy = 0. \end{cases}$

It is often possible to apply certain artificial methods of solving systems of equations, corresponding to special forms of the equations. Let us explain some of these methods by examples.

Example 1. Let the equations $x + y = 8$ and $xy = 15$ be given. The form of these equations shows that x and y can be regarded as the roots of a single quadratic equation $z^2 - 8z + 15 = 0$. The roots of the latter are 3 and 5. Since each of these roots may be taken as x and each as y, the given system has two systems of solutions: $x_1 = 3, y_1 = 5$ and $x_2 = 5, y_2 = 3$.

Example 2. Take the equations $x + y = 7$ and $x^2 + y^2 = 25$. Squaring the first and then subtracting the second, we find the product $xy = 12$. Knowing the sum and product of the unknowns, we can determine the unknowns as shown in the first example.

Example 3. Let the equations $x^2 - y^2 = 24$ and $x - y = 4$ be given. Dividing the first by the second, we find the first-degree equation $x + y = 6$, which together with the second of the given equations determines the unique system $x_1 = 5$ and $y_1 = 1$.

Example 4. Given the equations $x^2 + y^2 = 25$ and $xy = 12$. Multiplying the second equation by 2, we add it to the first and then subtract it from the first. We obtain $(x+y)^2 = 49$ and $(x-y)^2 = 1$, whence $x+y = \pm 7$ and $x-y = \pm 1$. Therefore the solutions of the given equations are obtained from the following systems of first-degree equations:

$$\begin{array}{llll} x + y = 7; & x + y = 7; & x + y = -7; & x + y = -7; \\ x - y = 1; & x - y = -1; & x - y = 1; & x - y = -1. \end{array}$$

These solutions are: $x_1 = 4, y_1 = 3$; $x_2 = 3, y_2 = 4$; $x_3 = -3, y_3 = -4$; $x_4 = -4, y_4 = -3$.

The same equations could be solved by means of a special substitution, which we explain in the following example.

Example 5. Take the equations $2xy - y^2 = 15$ and $x^2 + xy = 36$, whose left sides are homogeneous expressions of the second degree. Put $y = ux$. We obtain $x^2(2u - u^2) = 15$ and $x^2(1 + u) = 36$. Hence, determining two expressions for x^2 and comparing them, we find the equation $\frac{15}{2u-u^2} = \frac{36}{1+u}$, or $12u^2 - 19u + 5 = 0$. The roots of this equation are $u_1 = \frac{5}{4}$ and $u_2 = \frac{1}{3}$. By the first root we compute $x^2 = \frac{36}{1+u} = 16$, i.e. $x = \pm 4$, and consequently $y = ux = \pm 5$; by the second root we find also $x^2 = 27$, i.e. $x = \pm 3\sqrt{3}$, and consequently $y = \pm\sqrt{3}$. In all we obtain four systems of solutions.

Solve the following systems by either ordinary or artificial methods:

31a. $\begin{cases} x + y = 12; \\ xy = 35. \end{cases}$ **31b.** $\begin{cases} x - y = 8; \\ xy = 20. \end{cases}$

32a. $\begin{cases} x^2 + y^2 = 13; \\ x^2 - y^2 = 5. \end{cases}$ **32b.** $\begin{cases} x^2 + 2y^2 = 33; \\ 2x^2 - y^2 = 46. \end{cases}$

33a. $\begin{cases} x^2 + y^2 = 74; \\ x + y = 12. \end{cases}$ **33b.** $\begin{cases} x^2 + y^2 = 34; \\ x - y = 2. \end{cases}$

34a. $\begin{cases} x^2 - y^2 = 32; \\ x - y = 4. \end{cases}$ **34b.** $\begin{cases} x^2 - y^2 = 120; \\ x + y = 20. \end{cases}$

35a. $\begin{cases} \frac{x+y}{x-y} = \frac{3}{2}; \\ xy = 80. \end{cases}$ **35b.** $\begin{cases} \frac{x-y}{x+y} = \frac{3}{7}; \\ xy = 10. \end{cases}$

36a. $\begin{cases} \frac{1}{x} + \frac{1}{y} = 1; \\ x + y = 4. \end{cases}$ **36b.** $\begin{cases} \frac{1}{y} - \frac{1}{x} = \frac{1}{6}; \\ x - y = 1. \end{cases}$

37a. $\begin{cases} x^2 + y^2 = 25; \\ xy = 12. \end{cases}$ **37b.** $\begin{cases} x^2 - y^2 = 5; \\ xy = 6. \end{cases}$

38a. $\begin{cases} x^2 - xy + y^2 = 43; \\ x - y = 1. \end{cases}$ **38b.** $\begin{cases} x^2 + xy + y^2 = 67; \\ x + y = 9. \end{cases}$

39a. $\begin{cases} \sqrt{\frac{x}{y}} - \sqrt{\frac{y}{x}} = \frac{3}{2}; \\ x - y = 6. \end{cases}$

39b. $\begin{cases} \sqrt{\frac{x}{y}} + \sqrt{\frac{y}{x}} = \frac{5}{2}; \\ x + y = 10. \end{cases}$

40a. $\begin{cases} \frac{x-y}{\sqrt{x}-\sqrt{y}} = 10; \\ \sqrt{xy} = 16. \end{cases}$

40b. $\begin{cases} \frac{x-y}{\sqrt{x}+\sqrt{y}} = 2; \\ \sqrt{xy} = 15. \end{cases}$

41a. $\begin{cases} x^2 - y = 7; \\ x^2 y = 18. \end{cases}$

41b. $\begin{cases} x + y^2 = 11; \\ xy^3 = 18. \end{cases}$

42a. $\begin{cases} x^3 - y^3 = 37; \\ x - y = 1. \end{cases}$

42b. $\begin{cases} x^3 + y^3 = 65; \\ x + y = 5. \end{cases}$

43a. $\begin{cases} \frac{x}{y} + \frac{y}{x} = \frac{10}{3}; \\ x^2 - y^2 = 8. \end{cases}$

43b. $\begin{cases} \frac{x}{y} - \frac{y}{x} = \frac{3}{2}; \\ x^2 + y^2 = 45. \end{cases}$

44a. $\begin{cases} \frac{x^2}{y} + \frac{y^2}{x} = 18; \\ x + y = 12. \end{cases}$

44b. $\begin{cases} \frac{x^2}{y} + \frac{y^2}{x} = \frac{21}{2}; \\ x - y = 3. \end{cases}$

45a. $\begin{cases} 4x^2 + 9y^2 = 45; \\ xy = 3. \end{cases}$

45b. $\begin{cases} 25x^2 - y^2 = 36; \\ xy = 16. \end{cases}$

46a. $\begin{cases} \frac{x^2+y^2}{xy} = \frac{5}{2}; \\ x^2 - y^2 = 3. \end{cases}$

46b. $\begin{cases} \frac{x^2-y^2}{xy} = \frac{5}{6}; \\ x^2 + y^2 = 13. \end{cases}$

47a. $\begin{cases} x^3 - y^3 = 19; \\ x^2 y - xy^2 = 6. \end{cases}$

47b. $\begin{cases} x^3 + y^3 = 152; \\ x^2 y + xy^2 = 120. \end{cases}$

48a. $\begin{cases} \frac{x+y}{x-y} + \frac{x-y}{x+y} = \frac{5}{2}; \\ x^2 + y^2 = 20. \end{cases}$

48b. $\begin{cases} \frac{x+y}{x-y} + \frac{x-y}{x+y} = \frac{10}{3}; \\ x^2 + y^2 = 45. \end{cases}$

49a. $\begin{cases} x\sqrt{\frac{x}{y}} - y\sqrt{\frac{y}{x}} = \frac{65}{6}; \\ x - y = 5. \end{cases}$

49b. $\begin{cases} x\sqrt{\frac{x}{y}} - y\sqrt{\frac{y}{x}} = 30; \\ x + y = 20. \end{cases}$

50a. $\begin{cases} x^2 + y^2 - xy = 61; \\ x + y - \sqrt{xy} = 7. \end{cases}$

50b. $\begin{cases} x^2 + y^2 + xy = 84; \\ x + y - \sqrt{xy} = 6. \end{cases}$

51a. $x + y = xy = x^2 + y^2.$

51b. $x - y = xy = x^2 + y^2.$

52a. $x - y = x^2 - y^2 = x^3 - y^3.$

52b. $x + y = x^2 + y^2 = x^3 + y^3.$

53a. $\begin{cases} x + y = 5; \\ x^4 + y^4 = 97. \end{cases}$

53b. $\begin{cases} x - y = 2; \\ x^4 + y^4 = 82. \end{cases}$

54a. $\begin{cases} x - y = 3; \\ x^5 - y^5 = 33. \end{cases}$

54b. $\begin{cases} x + y = 2; \\ x^5 + y^5 = 242. \end{cases}$

55a. $\begin{cases} \frac{x^2}{y^2} + \frac{y^2}{x^2} + \frac{x}{y} + \frac{y}{x} = \frac{112}{9}; \\ x + y = 4. \end{cases}$

55b. $\begin{cases} \frac{x^2}{y^2} + \frac{y^2}{x^2} + \frac{x}{y} + \frac{y}{x} = \frac{27}{4}; \\ x - y = 2. \end{cases}$

56a. $\begin{cases} \frac{x^2}{y^2} + \frac{y^2}{x^2} + \frac{x}{y} - \frac{y}{x} = \frac{23}{4}; \\ x - y = 1. \end{cases}$

56b. $\begin{cases} \frac{x^2}{y^2} + \frac{y^2}{x^2} - \frac{x}{y} + \frac{y}{x} = \frac{11}{4}; \\ x + y = 3. \end{cases}$

57a. $\begin{cases} \sqrt{\frac{3x-2y}{2x}} + \sqrt{\frac{2x}{3x-2y}} = 2; \\ x^2 - 8 = 2x(2y - 3). \end{cases}$

57b. $\begin{cases} \sqrt{\frac{4x+3y}{5y}} + \sqrt{\frac{5y}{4x+3y}} = 2; \\ y^2 + 8 = 2y(x + 2). \end{cases}$

58a. $\begin{cases} \sqrt{\frac{6x}{x+y}} + \sqrt{\frac{x+y}{6x}} = \frac{5}{2}; \\ xy - x - y = 9. \end{cases}$

58b. $\begin{cases} \sqrt{\frac{5x}{x-y}} - \sqrt{\frac{x-y}{5x}} = \frac{21}{10}; \\ xy + x + y = 11. \end{cases}$

59a. $\begin{cases} x - y + \sqrt{\frac{x-y}{x+y}} = \frac{20}{x+y}; \\ x^2 + y^2 = 34. \end{cases}$

59b. $\begin{cases} x + y - \sqrt{\frac{x+y}{x-y}} = \frac{30}{x-y}; \\ xy = 80. \end{cases}$

60a.
$\begin{cases} x + y = 444; \\ \sqrt[3]{x+10} + \sqrt[3]{y+14} = 12. \end{cases}$

60b. $\begin{cases} x - y = 2; \\ \sqrt[3]{x+14} - \sqrt[3]{y-21} = 1. \end{cases}$

61a. $\begin{cases} xy = 12; \\ xz = 6; \\ y^2 + z^2 = 20. \end{cases}$

61b. $\begin{cases} xy = 54; \\ yz = 36; \\ x^2 - z^2 = 20. \end{cases}$

62a. $\begin{cases} xy = 48; \\ yz = 54; \\ xz = 72. \end{cases}$

62b. $\begin{cases} xy = 9z; \\ xz = 4y; \\ yz = 16x. \end{cases}$

63a. $\begin{cases} xy + yz = 28; \\ xz + yz = 30; \\ xy + xz = 10. \end{cases}$

63b. $\begin{cases} x^2 + y^2 = 52; \\ y^2 + z^2 = 100; \\ x^2 + z^2 = 80. \end{cases}$

64a. $\begin{cases} xy + xz + yz = 27; \\ x - y = 6; \\ y - z = 3. \end{cases}$

64b. $\begin{cases} x^2 + y^2 + z^2 = 98; \\ x - y = 5; \\ x + z = 8. \end{cases}$

65a. $\begin{cases} x(x + y + z) = 70; \\ y(x + y + z) = 28; \\ z(x + y + z) = 98. \end{cases}$

65b. $\begin{cases} x(x - y + z) = 12; \\ y(x - y + z) = 9; \\ z(x - y + z) = 6. \end{cases}$

66a. $\begin{cases} x + y + z = 20; \\ xyz = 130; \\ x - 2y + z = 5. \end{cases}$

66b. $\begin{cases} x - y + z = 8; \\ x^2 + y^2 + z^2 = 74; \\ x - y + 3z = 22. \end{cases}$

67a. $\begin{cases} x + y + z = 12; \\ xz + yz = 35; \\ x^2 + y^2 + z^2 = 50. \end{cases}$

67b. $\begin{cases} x - y + z = 3; \\ xz - yz = 2; \\ x^2 - y^2 + z^2 = 25. \end{cases}$

68a. $\begin{cases} x + y + z = 7; \\ x^2 + y^2 + z^2 = 21; \\ yz = x^2. \end{cases}$

68b. $\begin{cases} x + y + z = 6; \\ x^2 + y^2 + z^2 = 14; \\ yz = 6. \end{cases}$

69a. $\begin{cases} x^2 + y^2 = z^2; \\ x + y + z = 30; \\ xy = 60. \end{cases}$

69b. $\begin{cases} y^2 + z^2 = x^2 - 6; \\ x + y + z = 8; \\ xz = 3. \end{cases}$

70a. $\begin{cases} x^2 + z^2 - y^2 = 1; \\ x + y + z = 3; \\ y^2 = xz. \end{cases}$

70b. $\begin{cases} x^2 + y^2 + z^2 = 35; \\ x - y + z = 3; \\ y^2 = xz + 4. \end{cases}$

71a. $\begin{cases} x+y+z=13; \\ x^2+y^2+z^2=61; \\ 2xz=xy+xz. \end{cases}$

71b. $\begin{cases} x-y+z=14; \\ x^2+y^2+z^2=244; \\ 2z(x-y)=xy. \end{cases}$

72a. $\begin{cases} x^2+y^2+z^2=30; \\ y^2=2xz+21; \\ 2x=z. \end{cases}$

72b. $\begin{cases} xy+xz-yz=14; \\ z^2=2xy-4; \\ 3x=2z. \end{cases}$

73a. $\begin{cases} \frac{1}{x}+\frac{1}{y}+\frac{1}{z}=12; \\ \frac{3}{x}+\frac{2}{y}=18; \\ 3y+10z=3. \end{cases}$

73b. $\begin{cases} \frac{1}{x}-\frac{1}{y}+\frac{1}{z}=6; \\ \frac{4}{x}-\frac{3}{y}=7; \\ 8x-5z=1. \end{cases}$

74a. $\begin{cases} \frac{1}{x}+\frac{1}{y}-\frac{1}{z}=5; \\ \frac{1}{x}+\frac{1}{z}=6; \\ \frac{3}{y}-\frac{1}{xz}=1. \end{cases}$

74b. $\begin{cases} \frac{1}{x}+\frac{1}{y}+\frac{1}{z}=13; \\ \frac{1}{y}-\frac{1}{x}=1; \\ \frac{1}{xy}-\frac{2}{z}=0. \end{cases}$

75a. $\begin{cases} x+y+z=6; \\ xy+xz+yz=11; \\ xyz=6. \end{cases}$

75b. $\begin{cases} x-y+z=0; \\ xz-xy-yz=-31; \\ xyz=30. \end{cases}$

76a. $\begin{cases} x+y+z=0; \\ xyz=30; \\ x^2+y^2+z^2=38. \end{cases}$

76b. $\begin{cases} x+y+z=9; \\ xyz=24; \\ x^2+y^2+z^2=29. \end{cases}$

77a. $\begin{cases} x+u=5; \\ y+z=9; \\ y^2+u=28; \\ x+z^2=18. \end{cases}$

77b. $\begin{cases} u-x=3; \\ z-y=5; \\ u+y^2=12; \\ z^2-x=44. \end{cases}$

78a. $\begin{cases} x+u=10; \\ y-z=1; \\ yz=20; \\ y^2+u^2=74. \end{cases}$

78b. $\begin{cases} u-x=5; \\ x^2+z^2=52; \\ xz=24; \\ y^2+u^2=90. \end{cases}$

79a. $\begin{cases} xu=yz; \\ x+u=13; \\ y+z=11; \\ x^2+y^2+z^2+u^2=170. \end{cases}$

79b. $\begin{cases} xy=zu; \\ x+y=11; \\ z-u=2; \\ x^2+y^2-z^2-u^2=21. \end{cases}$

80a. $\begin{cases} x^3+y^3+z^3+u^3=252; \\ x+y=5; \\ z+u=7; \\ xy=zu. \end{cases}$

80b. $\begin{cases} x^3+y^3-z^3+u^3=187; \\ x+y=8; \\ z-u=1; \\ xy=zu. \end{cases}$

81a. $\begin{cases} \sqrt{x}-\sqrt{y}=a; \\ \sqrt{xy}=b. \end{cases}$

81b. $\begin{cases} \frac{\sqrt{x}+\sqrt{y}}{\sqrt{x}-\sqrt{y}}=\frac{a}{b}; \\ xy=(a^2-b^2)^2. \end{cases}$

82a. $\begin{cases} x^3+xy^2=a; \\ y^3+x^2y=b. \end{cases}$

82b. $\begin{cases} ax^2+bxy=a; \\ by^2+axy=b. \end{cases}$

83a. $\begin{cases} x^2+y^2=5a^2; \\ y^2+z^2=13a^2; \\ z^2-8x^2=a^2. \end{cases}$

83b. $\begin{cases} x^2-y^2=4an; \\ y^2-z^2=n^2-2an; \\ x^2+z^2=n^2+2an+2a^2. \end{cases}$

84a. $\begin{cases} \frac{x^2}{a}=\frac{y^2}{b}=\frac{z^2}{c}; \\ bcx^2+acy^2+abz^2=m^2. \end{cases}$

84b. $\begin{cases} x^2+y^2-z^2=a^2; \\ \frac{x}{m}=\frac{y}{n}=\frac{z}{p}. \end{cases}$

85a. $\begin{cases} xy=a; \\ xz=b; \\ y^2+z^2=n^2. \end{cases}$

85b. $\begin{cases} x^2+y^2=a; \\ x^2+z^2=b; \\ y^2+z^2=c. \end{cases}$

86a. $\begin{cases} x(y+z)=a; \\ y(x+z)=b; \\ z(x+y)=c. \end{cases}$

86b. $\begin{cases} x+y=az; \\ x-y=bz; \\ x^2+y^2=cz. \end{cases}$

87a. $\begin{cases} x^2-(y-z)^2=a; \\ y^2-(x-z)^2=b; \\ z^2-(x-y)^2=c. \end{cases}$

87b. $\begin{cases} x(y+z)=4a^2; \\ y(x+z)=3a^2; \\ z(x+y)=2a^2. \end{cases}$

88a. $\begin{cases} x^2+y^2=a; \\ y^2+z^2=b; \\ z^2+x^2=c. \end{cases}$

88b. $\begin{cases} (x+y)(x+z)=a^2; \\ (x+y)(y+z)=b^2; \\ (x+z)(y+z)=c^2. \end{cases}$

89a. $\begin{cases} x+y=a; \\ x+z=b; \\ x^2=y^2+z^2. \end{cases}$

89b. $\begin{cases} x+y+z=a; \\ yz=bx; \\ x^2=y^2+z^2. \end{cases}$

90a. $\begin{cases} 2x-3y+z=5n; \\ x^2+y^2+z^2=3m^2+2n^2; \\ xy+xz+yz=3m^2-n^2. \end{cases}$

90b. $\begin{cases} x+y+z=2m+n; \\ x^2+y^2+z^2=2m^2+3n^2; \\ (x+y)z=2mn. \end{cases}$

Composing Systems of Equations

91. Decompose the number 209 into two factors whose sum equals 30.

92. Decompose the number 195 into two factors whose difference equals 2.

93. The sum of two numbers is 13, the sum of their squares is 89. Find these numbers.

94. The difference of the squares of two numbers is 200. If each of the numbers is decreased by 1, the difference of their squares becomes 180. Find these numbers.

95. If to the product of two numbers one adds the smaller of them, one obtains 54. If to the same product one adds the larger number, one obtains 56. Find these numbers.

96. The product of the digits of a two-digit number is one third of the number itself. If 18 is added to the sought number, one obtains a number with the same digits but in reverse order. Find the number.

97. The product of two positive integers is 3 times their sum, and the sum of the squares of the same numbers equals 160. Find these numbers.

98. The sum of two numbers is 22, and the sum of their cubes is 2926. Find these numbers.

99. The difference of two numbers is 3, and the difference of their cubes is 657. Find these numbers.

100. Find a fraction such that the sum of the squares of its elements equals 25 and the sum of this fraction with its reciprocal equals $\frac{25}{12}$.

101. The sum of the squares of the digits of a two-digit number is 34; the product of the sought number by its reverse equals 1855. Find the number.

102. The sum of three numbers forming a continued arithmetic (difference) proportion equals 54, and their product equals 5760. Find these numbers.

103. The sum of three numbers forming a continued arithmetic (difference) proportion equals 12, and the sum of their squares equals 66. Find these numbers.

104. The sum of the digits of a three-digit number is 11; the sum of the squares of the same digits 45. If 198 is subtracted from the sought number, one obtains the reversed number. Find this number.

105. The sum of three numbers forming a continued geometric (multiple) proportion equals 19, and the sum of their squares 133. Find these numbers.

106. Find a three-digit number from the following conditions: the quotient of the sought number divided by the sum of its digits is 48; the quotient of the product of the digits divided by the same sum is $10\frac{2}{3}$; the tens digit is the arithmetic mean of the remaining digits.

107. Find two numbers the sum of whose squares exceeds the sum of their first powers by the quantity a, and the difference of whose squares exceeds the difference of their first powers by the quantity b.

108. The ratio of the sum of two numbers to their difference equals the ratio of p to q, and the product of the sum of the same numbers by their difference equals a. Find these numbers.

109. The area of a rectangle is 112 cm^2. The sum of the areas of the squares constructed on the adjacent sides of the rectangle is 260 cm^2. Find the sides of the rectangle.

110. The ratio of the sides of a rectangle is 6. The sum of the areas of the squares constructed on these sides is 592 cm^2. Find the sides of the rectangle.

111. The height of a trapezoid is 18 cm. Its area is equal to the area of a rectangle constructed on the base of the trapezoid; triple the upper base, added to the lower, is 4 times the height. Determine the bases of the trapezoid.

112. The hypotenuse of a right triangle is 1 cm greater than a leg, and the sum of the hypotenuse with this leg is 5 times the other leg. Determine the sides of this triangle.

113. The perimeter of a right triangle is 24 cm, the area 24 cm^2. Find the sides of the triangle.

114. The perimeter of a right triangle is 30 cm, the area 30 cm^2. Find the sides of the triangle.

115. The sum of the diagonals of a rhombus is 6 cm less than its perimeter. Compute the length of the side of the rhombus and its diagonals, if the area of the rhombus equals 24 cm^2.

116. The sum of the areas of two circles touching externally equals 90π cm^2. Compute the diameters of the circles, if the distance between their centers equals 12 cm.

117. Determine the legs of a right triangle whose hypotenuse equals a and area s.

118. Determine the sides of a right triangle from its given area s and perimeter $2p$.

119. Find the sides of a right triangle from the perimeter $2p$ and the difference of the legs d.

120. Find the diagonals of a rhombus from its side a and area s.

121. A body at rest begins to move with uniform acceleration. Having travelled a path of 1440 m, it acquired a speed of 24 m per second. How long, and with what acceleration, did the body move?

122. The resultant of two forces intersecting at a right angle equals 15 kg. If one of these forces is decreased by 4 kg and the other increased by 6 kg, their resultant becomes 17 kg. Find the magnitudes of the acting forces.

123. Between two cities at a distance of 450 km, a passenger train and a freight train run. The passenger covers the whole distance 8 hours faster than the freight. After increasing the speed of both trains by 5 km per hour, the passenger train came to cover the whole distance only 6 hours faster than the freight. Find the speed of both trains.

124. Two cyclists set out simultaneously toward each other from cities A and B, the distance between which is 28 km. After 1 hour of riding they met and, without stopping, continued on their way. The first arrived in city B 35 minutes sooner than the second arrived in city A. Find the speed of each, and also the distance each travelled before meeting.

125. Two bodies begin to move uniformly along two straight lines intersecting at a right angle, toward their point of intersection. At the beginning of motion the distance between the bodies is 169 cm, the speed of the first body 1 cm/sec, of the second 4 cm/sec. Determine the distances of the bodies to the point of intersection of their paths, given that 2 seconds after the start of their motion the distance between the bodies will be 25 cm.

126. Two pipes are led into a tank; through the first water flows out, through the second it flows in. With the joint action of both pipes the tank fills in 24 hours. If the cross-sectional areas of the pipes were increased so that the first pipe emptied the tank two hours faster, and the second also filled it two hours faster, then with the joint action of both pipes the tank would fill in 12 hours.

In how many hours does the first pipe empty the tank, and in how many hours does the second fill it?

127. Over a stretch of 18 m the front wheel of a carriage makes 10 more revolutions than the rear. If the circumference of the front wheel were increased by 6 dm and the circumference of the rear decreased by 6 dm, then over the same stretch the front wheel would make 4 more revolutions than the rear. Find the circumference of both wheels.

128. Over a stretch of 27 m the rear wheel of a carriage makes 5 fewer revolutions than the front. If the circumference of the rear wheel were increased by 3 dm and the circumference of the front decreased by 3 dm, then over the same stretch the rear wheel would make 9 fewer revolutions than the front. Find the circumference of both wheels.

129. A body was weighed on a faulty balance, i.e. one with unequal arms; when placing the weights on one pan the weight came out p grams, on the other q grams. Find the true weight of the body.

130. Three conductors connected in series have a resistance of 11 ohms. When connected in parallel their resistance equals 1 ohm. Compute the resistance of each of the conductors, if the resistance of one of them is 1 ohm less than the sum of the resistances of the other two.

131. Determine the specific gravity of two bodies, if a mixture containing a kilograms of the first body and b kilograms of the second has specific gravity p, and a mixture containing c kilograms of the first and d kilograms of the second has specific gravity q.

132. From two points M and N at a distance of d centimeters, two bodies begin to move simultaneously, which meet when the first body has travelled a centimeters. Determine the speeds of the bodies, given that the number of centimeters expressing the difference between the speeds of the first and second bodies equals the number of seconds elapsed from the start of motion to the meeting.

133. Two bodies move along the circumference of a circle; the first runs around the circumference d seconds faster than the second, and if they move in the same direction they converge every m seconds. What fraction of the circumference does each body run per second?

134. Three bodies move along a straight line from point M to point N. The second body began to move a seconds, and the third b seconds, later than the first. The speed of the first body is less than the speed of the second by c centimeters, the speed of the third equals v centimeters per second. Find the distance MN and the speed of the first body, given that all the bodies reach point M at one and the same moment.

Graphical Solution of Systems of Two Equations of the Second Degree in Two Unknowns

Having drawn the graphs of each of the equations of the system, one determines the coordinates of the points of intersection of these graphs; the values of the coordinates give the sought roots of the system of equations.

Solve the following systems of equations graphically:

135. $\begin{cases} x + y = 5; \\ xy = 2. \end{cases}$

136. $\begin{cases} x - y = 2; \\ xy = 3. \end{cases}$

137. $\begin{cases} x^2 + y^2 = 25; \\ y = \frac{4}{3}x. \end{cases}$

138. $\begin{cases} x^2 + y^2 = 41; \\ x - y = 1. \end{cases}$

139. $\begin{cases} x^2 + y^2 = 61; \\ xy = 30. \end{cases}$

140. $\begin{cases} x + y^2 = 19; \\ y + x^2 = 13. \end{cases}$

Chapter XV

Progressions

The Arithmetic, or Difference, Progression

An *arithmetic*, or *difference*, progression is a sequence of numbers in which each term, beginning with the second, is obtained from the preceding one by adding to it one and the same constant number, called the *difference* of the arithmetic progression (denoted by the letter d). Any term of the progression may be taken as the first term; it is denoted a_1 (or simply a), and the following ones a_2, a_3, $a_4 \dots a_n$ (or simply u). If the difference d is positive, the progression is called increasing; if the difference is negative, decreasing. The sum of some number of terms of an arithmetic progression, from some a_1 taken as the first to some a_n taken as the last, is denoted by s_n (or simply s). The formulas of the arithmetic progression are:

$$a_n = a_1 + d(n-1), \text{ or } u = a + d(n-1);$$

$$s_n = \frac{a_1 + a_n}{2} \cdot n, \text{ or } s = \frac{a+u}{2} \cdot n.$$

It is useful to know one more formula for the sum of the terms of an arithmetic progression:

$$s = \frac{2a + d(n-1)}{2} \cdot n.$$

1a. Find the 15th term and the sum of 15 terms of the progression 2, 5, 8, 11...

2a. Find the 18th term and the sum of 18 terms of the progression −3, −5, −7, −9...

3a. Find the sum of all two-digit numbers from 21 to 50 inclusive.

4a. Find the sum of the series of consecutive integers from 1 to n.

5a. Find the sum of the natural numbers from 1 to 100.

1b. Find the 20th term and the sum of 20 terms of the progression 3, 7, 11, 15...

2b. Find the 13th term and the sum of 13 terms of the progression −2, −6, −10, −14...

3b. Find the sum of all two-digit numbers.

4b. Find the sum of the series of consecutive integers from 1 to $2n$.

5b. Find the sum of the natural numbers from 1 to 1000.

6a. Find the sum of all even numbers up to 200 inclusive.

7a. Find the n-th odd number and the sum of n odd numbers.

8a. Find the sum of n terms of the progression $a, 2a - b, 3a - 2b \ldots$

9a. Between the numbers 3 and 24 insert 6 arithmetic means, i.e. so that the sought numbers together with the given ones form an arithmetic progression.

10a. Between 27 and -28 insert 10 arithmetic means.

11a. $a = 7$; $d = 4$; $n = 13$. Find u and s.

12a. $a = 56$; $d = -3$; $n = 11$. Find u and s.

13a. $u = 149$; $d = 7$; $n = 22$. Find a and s.

14a. $u = -22$; $d = -2$; $n = 40$. Find a and s.

15a. $a = 2$; $u = 87$; $s = 801$. Find d and n.

16a. $a = 10$; $u = -9$; $s = 10$. Find d and n.

17a. $a = 3$; $u = 63$; $n = 16$. Find d and s.

18a. $a = 36$; $u = 8$; $n = 15$. Find d and s.

19a. $a = 10$; $n = 14$; $s = 1050$. Find u and d.

20a. $a = -45$; $n = 31$; $s = 0$. Find u and d.

21a. $u = 21$; $n = 7$; $s = 105$. Find a and d.

22a. $u = 105$; $n = 16$; $s = 840$. Find a and d.

23a. $a = 4$; $d = 5$; $u = 49$. Find n and s.

24a. $a = 14.5$; $d = 0.7$; $u = 32$. Find n and s.

25a. $d = 6$; $n = 10$; $s = 340$. Find a and u.

26a. $d = \frac{1}{2}$; $n = 25$; $s = -75$. Find a and u.

27a. $a = 2$; $d = 5$; $s = 245$. Find n and u.

6b. Find the sum of all odd numbers up to 175 inclusive.

7b. Find the n-th even number and the sum of n even numbers.

8b. Find the sum of n terms of the progression $b, 2b - a, 3b - 2a \ldots$

9b. Between the numbers 17 and 82 insert 12 arithmetic means.

10b. Between 17 and -19 insert 17 arithmetic means.

11b. $a = 2$; $d = 2$; $n = 40$. Find u and s.

12b. $a = 63$; $d = -5$; $n = 8$. Find u and s.

13b. $u = 65$; $d = 5$; $n = 12$. Find a and s.

14b. $u = 13$; $d = -3$; $n = 58$. Find a and s.

15b. $a = -13$; $u = 27$; $s = 77$. Find d and n.

16b. $a = 160$; $u = 17$; $s = 1062$. Find d and n.

17b. $a = 1$; $u = 81$; $n = 17$. Find d and s.

18b. $a = 169$; $u = 8$; $n = 24$. Find d and s.

19b. $a = -40$; $n = 20$; $s = -40$. Find u and d.

20b. $a = 16$; $n = 9$; $s = 0$. Find u and d.

21b. $u = 92$; $n = 11$; $s = 517$. Find a and d.

22b. $u = -143$; $n = 33$; $s = -2079$. Find a and d.

23b. $a = 1$; $d = 3$; $u = 22$. Find n and s.

24b. $a = -28$; $d = 7$; $u = 28$. Find n and s.

25b. $d = \frac{1}{3}$; $n = 50$; $s = 425$. Find a and u.

26b. $d = -\frac{3}{4}$; $n = 33$; $s = -33$. Find a and u.

27b. $a = 40$; $d = -4$; $s = 180$. Find n and u.

28a. $a = 41$; $d = 2$; $s = 4784$. Find n and u.

28b. $a = 18$; $d = 6$; $s = 1782$. Find n and u.

29a. $d = 3$; $u = 29$; $s = 155$. Find n and a.

29b. $d = 5$; $u = 77$; $s = 623$. Find n and a.

30a. $d = 4$; $u = 88$; $s = 1008$. Find n and a.

30b. $d = 1.5$; $u = 45$; $s = 682.5$. Find n and a.

31a. The third term of a progression is 25, the tenth is -3. Find the first term and the difference.

31b. The fifth term of a progression is 13, the ninth 19. Find the first term and the difference.

32a. The fourth term of a progression is 10, the seventh 19. Find the sum of ten terms.

32b. The fifth term of a progression is -8, the seventeenth 28. Find the sum of fifteen terms.

33a. The fourth term of a progression is 9, the ninth -6. How many terms must be taken so that their sum equals 54?

33b. The tenth term of a progression is 4, the nineteenth -32. How many terms must be taken so that their sum equals 180?

34a. The sum of the third and seventh terms of a progression equals 4, and the sum of the second and fourteenth equals -8. Find the progression.

34b. The sum of the fourth and tenth terms of a progression equals 44, and the sum of the second and fifteenth 53. Find the progression.

35a. Find the difference of a progression whose first term equals 100 and the sum of whose first six terms is 5 times the sum of the next six terms.

35b. Find the first term of a progression whose difference equals 4 and the sum of whose first five terms is one third of the sum of the next five terms.

36a. Form a progression from 1 to 21 such that the sum of all its terms is to the sum of the terms between 1 and 21 as $11 : 9$.

36b. Form a progression from 1 to 29 such that the sum of all its terms is to the sum of the terms between 1 and 29 as $4 : 3$.

37a. The first term of a progression is 1; the sum of its first m terms is to the sum of its first n terms as $m^2 : n^2$. Find the progression.

37b. The first term of a progression is 2; the sum of its first m terms is to the sum of its first n terms as $m(m+1) : n(n+1)$. Find the progression.

38a. Find the sum of $m + n$ terms of a progression in which the m-th term equals n and the n-th term equals m.

38b. Find the sum of $m - n$ terms of a progression in which the sum of m terms equals n and the sum of n terms equals m.

39a. Find the progression, given that the sum of its first and second terms equals 7 and their product equals 10.

39b. Find the progression, given that the difference between its second and first terms equals a and their product equals $\frac{a^2}{16}$.

40. Find the sum of 100 terms of the series $1 - 3 + 5 - 7 + \ldots$

41a. Find the progression, given that the sum of its first three terms equals 15 and their product 80.

41b. Find the progression, given that the sum of its first three terms equals zero and the sum of their squares 50.

42a. Find the progression, given that the sum of its second and fourth terms equals 16 and the product of the first term by the fifth equals 28.

42b. Find the progression, given that the sum of its first and fifth terms equals 12 and the product of the second term by the fourth equals 32.

43. The numbers of degrees contained in the consecutive interior angles of a polygon form a progression with difference 10°; the smallest angle of this polygon is 100°. How many sides does the polygon have?

44. The numbers of degrees contained in the consecutive interior angles of a polygon form a progression with difference 5°; the smallest angle of this polygon is 120°. How many sides does the polygon have?

45. A freely falling body in airless space travels 4.9 m in the first second, and in each following second 9.8 m more than in the preceding. What distance will it travel in the 21st second and in 21 seconds?

46. How many seconds will a body fall from a height of 4410 m?

47. A body thrown upward loses 9.8 m of its speed each second. How long will a body thrown vertically with a speed of 656.6 m per second fly upward?

48. After what time will a body thrown vertically upward with a speed of 784 m per second fall to the ground?

49. With what speed was a body thrown vertically upward if it fell back to the ground after 1.5 minutes?

50. To what height did a vertically thrown body rise, and with what speed was it thrown, if after a minute it fell back to the ground?

51. Between the sides of an angle, 15 parallel segments are drawn at equal distances from one another, each following one being 3 cm longer than the preceding. Find the length of the first and of the last, if the sum of all the segments is 405 cm.

52. In mountainous regions the air temperature in summer drops on average by 0.7°C for each 100 m of ascent. At 11 o'clock a thermometer on a mountain showed 14.8°; how high is the observer, if at the foot at that time the temperature was 26°C?

53. After how many seconds will a stone fall into a well 44.1 m deep (air resistance not taken into account)?

54. A stepped pulley consists of six steps. Their diameters form an arithmetic progression. The largest diameter equals 240 mm, the smallest 80 mm. Find the remaining diameters (Fig. 4).

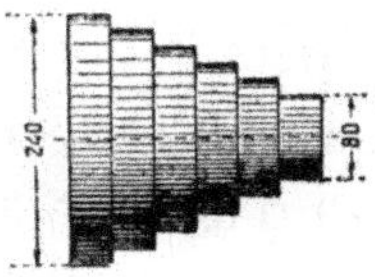

Fig. 4.

55. A tourist, climbing a mountain, rose 900 m on the first day, and on each following day continued to rise to a height 50 m less than on the preceding. In how many days did he reach the summit of the mountain, if its height is 5250 m?

56. A locomotive arrived from the passenger terminal at the freight station, the distance between which is 3 km. In doing so, in the first minute it travelled 480 m, and in each following one travelled 40 m less than in the preceding. In how many minutes did it make the run?

57. Along a highway a carriage and an automobile travel in the same direction. The carriage is 84 m ahead of the automobile and travels uniformly at a speed of 3 m per second. The automobile travels 8 m in the first second, and in each following one 0.1 m less than in the preceding. How long will the automobile travel until it draws level with the carriage?

58. Two automobiles move toward each other from two places, the distance between which equals 240 m. The first automobile set out 3 seconds earlier than the second and moves uniformly at a speed of 10 m per second. The second travels 2 m in the first second of its motion, and in each following one 1 m more than in the preceding. After how many seconds from the departure of the first automobile will their meeting occur?

59. It is supposed that for each 30.5 m of descent the internal temperature of the earth increases by 1°C. If at the surface of the earth the temperature equals 10°C, then at what depth does it reach: the boiling point of water (100°), the melting of lead (334°), the melting of iron (1200°)? Assuming the stated law does not change up to the very center of the earth, find the temperature at the center of the earth. The mean radius of the earth equals 6370 km.

60. A school cooperative acquired educational supplies for a sum of 800 rubles, on condition of paying this sum over several months so that in the first month it pays 120 rubles, and in each following month pays 20 rubles more than in the preceding month. After how many months will the cooperative pay off the entire stated sum?

61. A stone dropped into a well reached the surface of the water after 4 seconds. Determine the depth of the well to the surface of the water.

62. An aviator, jumping from the basket of a tethered balloon, flies down 100 m before the parachute opens. How much time passes until the parachute opens?

63. Several prizes are designated for a shooting competition; the largest of them is 150 rubles, and the value of each following prize decreases by one and the same amount down to the smallest of 30 rubles. The total value of the prizes is 360 rubles. How many prizes in all are designated?

64. The perimeter of a polygon equals 158 cm, the lengths of its sides forming an arithmetic progression with difference 3 cm. The largest side of the polygon equals 44 cm. How many sides does the polygon have?

65. A train, departing from a station, uniformly increases its speed and after 26 minutes reaches a speed of 60 km per hour. What is the acceleration of the train per minute?

The Geometric, or Multiple, Progression

A *geometric*, or *multiple*, progression is a sequence of numbers in which each term, beginning with the second, is obtained from the preceding by multiplying it by one and the same constant number, called the *ratio* of the geometric progression (denoted by the letter q). If the ratio q is greater than 1, the progression is called increasing; if less than 1, decreasing; if it is negative, the progression is an alternating series. The formulas of the geometric progression are:

$$a_n = a_1 q^{n-1}, \text{ or } u = aq^{n-1};$$

$$s_n = \frac{a_n q - a}{q-1}, \text{ or } s = \frac{uq-a}{q-1}, \text{ or } s = \frac{a-uq}{1-q}.$$

66a. Find the sum of 10 terms of the progression $10, 20, 40\ldots$

67a. Find the sum of 7 terms of the progression $-4, 16, -64\ldots$

68a. Find the sum of 8 terms of the progression $3, -1, \frac{1}{3}\ldots$

69a. Find the sum of 5 terms of the progression $\sqrt{\frac{2}{3}}, 1, \sqrt{\frac{3}{2}}\ldots$

70a. Find the sum of n terms of the progression $\frac{2}{3}, \frac{1}{2}, \frac{3}{8}\ldots$

71a. Find the sum of n terms of the progression $\sqrt{6}, 3\sqrt{2}, 3\sqrt{6}\ldots$

72a. Between the numbers 47 and 1269 insert 2 mean proportional quantities.

73a. Between the numbers $\frac{a}{b^2}$ and $\frac{b}{a^2}$ insert 5 mean proportional quantities.

74a. Find the sum of 6 terms of a geometric progression, any term of which is expressed in the form $3 \cdot 2^{n-1}$, where n is the order (number) of the term.

66b. Find the sum of 8 terms of the progression $5, 15, 45\ldots$

67b. Find the sum of 10 terms of the progression $3, 6, 12\ldots$

68b. Find the sum of 11 terms of the progression $2, 1, \frac{1}{2}\ldots$

69b. Find the sum of 7 terms of the progression $\sqrt{\frac{5}{6}}, 1, \sqrt{\frac{6}{5}}\ldots$

70b. Find the sum of n terms of the progression $\frac{3}{4}, \frac{1}{2}, \frac{1}{3}\ldots$

71b. Find the sum of n terms of the progression $\frac{1}{2}, \frac{1}{\sqrt{2}}, 1\ldots$

72b. Between the numbers 31 and 496 insert 3 mean proportional quantities.

73b. Between the numbers $\frac{b^2}{a^3}$ and $\frac{a^2}{b^3}$ insert 9 mean proportional quantities.

74b. Find the sum of 5 terms of a progression, any term of which is expressed in the form $2 \cdot 5^{m-1}$.

75a. Find the sum of n terms of a progression, any term of which is expressed in the form $(-1)^n a^{n-1} b^{m-n+1}$.

76a. $a = 6$; $q = 3$; $n = 8$. Find u and s.

77a. $a = 5$; $q = -\frac{1}{5}$; $n = 6$. Find u and s.

78a. $u = 128$; $q = 2$; $n = 7$. Find a and s.

79a. $u = \frac{2}{27}$; $q = -\frac{2}{3}$; $n = 5$. Find a and s.

80a. $a = 3$; $u = 12288$; $n = 5$. Find q and s.

81a. $a = 81$; $u = -10\frac{2}{3}$; $n = 6$. Find q and s.

82a. $q = 2$; $n = 7$; $s = 635$. Find a and u.

83a. $q = -\frac{1}{2}$; $n = 8$; $s = \frac{85}{16}$. Find a and u.

84a. $a = 3$; $q = 2$; $u = 96$. Find n and s.

85a. $a = 9$; $q = \frac{2}{3}$; $u = \frac{32}{27}$. Find n and s.

86a. $a = 2$; $u = 1458$; $s = 2186$. Find q and n.

87a. $a = 3$; $u = 96$; $s = 189$. Find q and n.

88a. $a = 7$; $q = 3$; $s = 847$. Find u and n.

89a. $a = 2$; $q = -3$; $s = -364$. Find u and n.

90a. $u = -216$; $q = -6$; $s = -186$. Find a and n.

91a. $u = 32768$; $q = 4$; $s = 43690$. Find a and n.

92a. $a = 15$; $n = 4$; $s = 225$. Find q and u.

93a. $a = 12$; $n = 3$; $s = 372$. Find q and u.

94a. $u = -\frac{32}{9}$; $n = 6$; $s = \frac{133}{9}$. Find q and a.

95a. $u = 135$; $n = 3$; $s = 195$. Find q and a.

75b. Find the sum of n terms of a progression, any term of which is expressed in the form $(-1)^n a^{m-n+1} b^{n-1}$.

76b. $a = -5$; $q = 2$; $n = 9$. Find u and s.

77b. $a = \frac{3}{4}$; $q = \frac{2}{3}$; $n = 10$. Find u and s.

78b. $u = 78125$; $q = 5$; $n = 8$. Find a and s.

79b. $u = -243$; $q = -\frac{3}{2}$; $n = 6$. Find a and s.

80b. $a = 8$; $u = 10368$; $n = 5$. Find q and s.

81b. $a = \frac{1}{64}$; $u = -\frac{16}{243}$; $n = 6$. Find q and s.

82b. $q = -2$; $n = 8$; $s = 85$. Find a and u.

83b. $q = \frac{1}{3}$; $n = 6$; $s = \frac{364}{9}$. Find a and u.

84b. $a = 5$; $q = 3$; $u = 405$. Find n and s.

85b. $a = \frac{3}{8}$; $q = -4$; $u = 96$. Find n and s.

86b. $a = 1$; $u = 2401$; $s = 2801$. Find q and n.

87b. $a = 2$; $u = 1458$; $s = 1514$. Find q and n.

88b. $a = 8$; $q = 2$; $s = 4088$. Find u and n.

89b. $a = 3$; $q = -2$; $s = 33$. Find u and n.

90b. $u = 250$; $q = 5$; $s = 312$. Find a and n.

91b. $u = 1215$; $q = -3$; $s = 915$. Find a and n.

92b. $a = 12$; $n = 4$; $s = 480$. Find q and u.

93b. $a = 15$; $n = 3$; $s = 105$. Find q and u.

94b. $u = -\frac{243}{2}$; $n = 6$; $s = -66.5$. Find q and a.

95b. $u = 8$; $n = 3$; $s = 14$. Find q and a.

96a. The first term of a progression equals 1; the sum of the third and fifth terms is 90. Find the progression.

97a. The sum of the first and third terms of a progression equals 15, and the sum of the second and fourth 30. Find the sum of ten terms.

98a. Find four numbers forming a geometric progression, given that the first is greater than the second by 36 and the third greater than the fourth by 4.

99a. Three numbers, the sum of which equals 114, can be regarded as three consecutive terms of a geometric progression or as the 1st, 4th, and 25th terms of an arithmetic progression. Find these numbers.

100a. Find a progression of six terms, given that the sum of the first three equals 112 and of the last three 14.

101a. Three numbers forming a geometric progression give a sum of 26; if to these numbers one adds respectively 1, 6, and 3, one obtains three numbers forming an arithmetic progression. Find these numbers.

102a. If from four unknown numbers forming an arithmetic progression one subtracts respectively 2, 7, 9, and 5, one obtains numbers forming a geometric progression. Find the terms of the arithmetic progression.

103a. Find the m-th and n-th terms of a progression in which the $(m+n)$-th term equals k and the $(m-n)$-th equals l.

96b. The first term of a progression equals 3; the difference between the seventh and fourth terms is 168. Find the progression.

97b. The difference between the third and first terms of a progression equals 24, and the difference between the fifth and first 624. Find the sum of six terms.

98b. Find four numbers forming a geometric progression, given that the sum of the extreme terms equals 27 and the sum of the middle ones 18.

99b. Three numbers, the sum of which is 124, are three consecutive terms of a geometric progression and at the same time the 3rd, 13th, and 15th terms of an arithmetic progression. Find these numbers.

100b. Find a progression of six terms, given that the sum of the terms in odd places equals 455 and the sum of the terms in even places equals 1365.

101b. Three numbers forming an arithmetic progression give a sum of 15; if to these numbers one adds respectively 1, 4, and 19, one obtains three numbers forming a geometric progression. Find these numbers.

102b. If from four unknown numbers forming a geometric progression one subtracts respectively 5, 6, 9, and 15, one obtains numbers forming an arithmetic progression. Find the terms of the geometric progression.

103b. Find the n-th and $(m+p)$-th terms of a progression in which the m-th term equals k and the p-th equals l.

104a. Prove that in every geometric progression the product of terms equidistant from the beginning and from the end is a constant quantity, equal to the product of the extreme terms.

104b. Derive the formula for the product of n terms of a geometric progression.

105a. Compute the product of five terms of a geometric progression whose first term is $\sqrt{2}$ and ratio $\frac{1}{2}$.

105b. Compute the product of n terms of a geometric progression whose first term is $\sqrt{a}$ and ratio $\frac{1}{2}$.

106. The permissible number of revolutions of a lathe spindle ranges between $n_1 = 11$ and $n_{11} = 350$ per minute. Besides these numbers of revolutions, nine others are possible: $n_2, n_3 \ldots, n_{10}$, forming with the given ones a geometric progression. Find these numbers of revolutions.

107. In a belt embracing a pulley, the tensions increase in geometric progression as the angle of embrace increases by one and the same amount. For angles of embrace of 15° and 90° the tensions are respectively 9.56 kg and 17.41 kg. Calculate the tension at every 15° and determine the smallest tension at an angle of embrace equal to 0°.

108. The numbers expressing the length, width, and height of a rectangular parallelepiped form a geometric progression; the area of the base of the parallelepiped equals 108 m^2, and the total surface 888 m^2. Determine the dimensions of the parallelepiped.

109. The numbers expressing the length, width, and height of a rectangular parallelepiped form a geometric progression; the volume of the parallelepiped equals 216 m^3, and the diagonal $\sqrt{364}$ m. Determine the dimensions of the parallelepiped.

110. After each stroke of the piston under the bell of an air pump, 0.83 of the previously present amount of air remains. Determine how great the air pressure under the bell will be after 15 strokes, if the initial pressure was equal to 760 mm.

111. A vessel contains 50 L of 80-percent alcohol. How many liters of pure alcohol will still be in the vessel if from it one pours out 1 L of liquid 20 times and each time adds 1 L of water?

112. A vessel contained 1250 L of 80-percent alcohol. From it a certain amount of liquid was taken 3 times and replaced with the same amount of water. After this, 125 L of pure alcohol was found in the vessel. What amount of liquid was taken from the vessel each time?

If the absolute value of the ratio of a progression is less than unity, one may consider in it an unbounded sequence of terms. It is known from algebra that for an infinitely decreasing progression $\lim_{n\to\infty} a_n = 0$, and consequently from the formula $s_n = \frac{a - a_n q}{1-q}$, as $n \to \infty$, one obtains the formula $s = \frac{a}{1-q}$ for the limit of the sum of the terms of an infinitely decreasing progression.

Determine the limit of the sum in the following infinitely decreasing geometric progressions:

113a. $1+\frac{1}{2}+\frac{1}{4}+\frac{1}{8}+\dots$

113b. $1+\frac{1}{3}+\frac{1}{9}+\frac{1}{27}+\dots$

114a. $1-\frac{1}{3}+\frac{1}{9}-\frac{1}{27}+\dots$

114b. $1-\frac{1}{2}+\frac{1}{4}-\frac{1}{8}+\dots$

115a. $\sqrt{\frac{3}{2}}+\sqrt{\frac{2}{3}}+\frac{2}{3}\sqrt{\frac{2}{3}}+\dots$

115b. $\sqrt{5}+\sqrt{\frac{5}{2}}+\frac{\sqrt{5}}{2}+\dots$

116a. $\frac{\sqrt{2}+1}{\sqrt{2}-1}-\frac{1}{2-\sqrt{2}}+\frac{1}{2}-\dots$

116b. $\frac{2+\sqrt{3}}{2-\sqrt{3}}-1+\frac{2-\sqrt{3}}{2+\sqrt{3}}-\dots$

117a. Form an infinitely decreasing progression in which each term is K times the sum of all the terms following it.

117b. Form an infinitely decreasing progression in which each term is one K-th of the sum of all the terms following it.

118a. A rectilinear segment AB is bisected at point C, then AC is bisected at point D, then CD at point E, DE at point F, EF at point G, and so on to infinity. Determine the limiting distance of the point of division from point A.

118b. A rectilinear segment AB is bisected at point C, then BC is bisected at D, then CD at E, DE at F, EF at G, and so on to infinity. Determine the limiting distance of the point of division from A.

119a. In a square whose side is a, another square is inscribed by bisecting the sides; in this square a new square is inscribed in the same way, and so on to infinity. Determine the limits to which the sums of the sides and of the areas of all the squares tend.

119b. In a regular triangle whose side is a, another regular triangle is inscribed by bisecting the sides. In this triangle a new triangle is inscribed in the same way, and so on to infinity. Determine the limits to which the sums of the sides and of the areas of all the triangles tend.

120a. Given a regular triangle whose side is a; from its three altitudes a new regular triangle is constructed; from the altitudes of this second triangle, yet another regular triangle, and so on. Determine the limits of those algebraic sums in one of which the perimeters, and in the other the areas, of the triangles are alternately addends and subtrahends.

120b. Given a square with diagonal a; the side of this square is taken as the diagonal of a second square; the side of the second as the diagonal of a new square, and so on. Determine the limits of those algebraic sums in one of which the perimeters, and in the other the areas, of the squares are alternately addends and subtrahends.

121a. In a circle of radius R a square is inscribed, in the square a circle is inscribed, in this circle a second square, and so on. Determine the limiting values of the sums of the areas of all the circles and of all the squares.

122a. On the side of an angle of $45°$ a point is taken at a distance a from the vertex. From this point a perpendicular is dropped to the second side, from the foot of this perpendicular a new perpendicular to the first side, and so on to infinity. Find the limit of the sum of the lengths of these perpendiculars.

121b. In a circle of radius R a regular triangle is inscribed, in this triangle a circle is inscribed, in the circle a regular triangle, and so on. Determine the limiting values of the sums of the areas of all the circles and of all the triangles.

122b. On the side of an angle of $60°$ a point is taken at a distance a from the vertex. From this point a perpendicular is dropped to the second side, from the foot of this perpendicular a new perpendicular to the first side, and so on to infinity. Find the limit of the sum of the lengths of these perpendiculars.

Chapter XVI

Logarithms

General Properties of Logarithms

The logarithm of a number N to the base a is the exponent of the power to which the base a must be raised to obtain the number N. This relation is denoted thus: $x = \log_a N$.

Proceeding from the definition of the logarithm, solve the following problems:

1a. What number has logarithm 3 to the base 2?

1b. What number has logarithm 2 to the base 3?

2. What number has logarithm $\frac{1}{3}$ to the base 8?

3a. To what base does the number 32 have logarithm 5?

3b. To what base does the number 81 have logarithm 4?

4a. To what base does the number 4 have logarithm $\frac{1}{3}$?

4b. To what base does the number 9 have logarithm $\frac{1}{2}$?

5a. What is the logarithm of the number 16 when the base equals 2?

5b. What is the logarithm of the number 27 when the base equals 3?

6a. What is the logarithm of the number 3 when the base equals 81?

6b. What is the logarithm of the number 7 when the base equals 49?

7a. To what base is $\log 16$ equal to 2?

7b. To what base is $\log 81$ equal to 2?

8a. Find x, knowing that $\log_4 x = 3$.

8b. Find x, knowing that $\log_5 x = 3$.

9a. What number has, to the base 5, logarithm -2?

9b. What number has, to the base 3, logarithm -3?

10a. Find the logarithm of $\frac{1}{8}$ to the base 2.

10b. Find the logarithm of $\frac{1}{81}$ to the base 3.

11a. Find the logarithms of the number 1024, taking as bases the numbers 2, 4, and 32.

11b. Find the logarithms of the number 729, taking as bases the numbers 3, 9, and 27.

12a. Find the logarithms of the number 81, taking as bases the numbers $\frac{1}{3}$, $\frac{1}{9}$, $\frac{1}{81}$.

13a. What number has logarithm -3 to the base 8?

14a. To what base is the logarithm of $\frac{1}{243}$ equal to -5?

15a. Find the logarithms of the fraction $\frac{1}{64}$, taking as bases the numbers 2, 4, and 8.

16a. Find the logarithms of the fraction $\frac{1}{729}$, taking as bases the numbers $\frac{1}{3}$, $\frac{1}{9}$, $\frac{1}{27}$.

17a. The base equals $\frac{3}{4}$; find the numbers whose logarithms are: 0, 1, -1, 2, -2, 3, -3.

18a. The base equals $2\frac{1}{2}$; find the logarithms of the numbers: $\frac{2}{5}$, $6\frac{1}{4}$, 1, $\frac{8}{125}$.

19a. For what bases does the number 125 have logarithms 3, 1, -3, -1?

20a. If the base of logarithms equals 0.5, what are the logarithms of the numbers 1, 4, 2, $\frac{1}{4}$, 8, $\frac{1}{8}$?

21a. What number has logarithm $\frac{3}{4}$ to the base 3?

22a. Between what integers do the logarithms of the numbers 3, 5, 10, 25, 100, and 500 to the base 2 lie?

23a. Between what negative integers do the logarithms of the fractions 0.02, 0.034, 0.005, 0.000675, 0.00009 to the base 10 lie?

24a. To what base does the number 5 have logarithm 2?

25a. Find the number whose logarithm to the base 8 equals $-\frac{3}{4}$.

26a. To what base does the number 7 have logarithm $-1\frac{1}{2}$?

12b. Find the logarithms of the number 256, taking as bases the numbers $\frac{1}{2}$, $\frac{1}{4}$, $\frac{1}{16}$.

13b. What number has logarithm -4 to the base 6?

14b. To what base is the logarithm of $\frac{1}{64}$ equal to -3?

15b. Find the logarithms of the fraction $\frac{1}{729}$, taking as bases the numbers 3, 9, and 27.

16b. Find the logarithms of the fraction $\frac{1}{512}$, taking as bases the numbers $\frac{1}{2}$, $\frac{1}{4}$, $\frac{1}{8}$.

17b. The base equals $1\frac{1}{2}$; find the numbers whose logarithms are 0, 1, -1, 2, -2, 3, -3, 4, -4.

18b. The base equals $\frac{3}{5}$; find the logarithms of the numbers $\frac{5}{3}$, $2\frac{7}{9}$, 1, $\frac{27}{125}$.

19b. For what bases does the number 343 have logarithms 3, -3, 1, -1?

20b. If the base of logarithms equals 0.2, what are the logarithms of the numbers 1, 25, 5, 0.04, 125, 0.008?

21b. What number has logarithm $\frac{2}{3}$ to the base 2?

22b. Between what integers do the logarithms of the numbers 5, 12, 862, 1613, and 11 111 to the base 10 lie?

23b. Between what negative integers do the logarithms of the number 597 to the base 0.1 lie?

24b. To what base does the number 3 have logarithm 2?

25b. Find the number whose logarithm to the base 25 equals $-\frac{2}{3}$.

26b. To what base does the number 5 have logarithm $-\frac{3}{4}$?

27a. The base of logarithms is -8; find the numbers whose logarithms are: -1, 3, -2, $\frac{1}{3}$, $-\frac{1}{3}$.

27b. The base of logarithms is -81; find the numbers whose logarithms are: 2, -1, -2, $\frac{1}{4}$, $-\frac{1}{2}$.

28a. Find the logarithms of the numbers $-\frac{8}{27}$, $\frac{4}{9}$, $5\frac{1}{16}$ to the base equal to $-\frac{2}{3}$.

28b. Find the logarithms of the numbers $-\frac{1}{4}$, -2, -32, 64 to the base equal to $-\frac{1}{8}$.

29a. What is the logarithm of $\sqrt[5]{9}$ to the bases 3, 81, $\frac{1}{9}$, $\frac{1}{81}$?

29b. What is the logarithm of $\sqrt[3]{49}$ to the bases 7, $\frac{1}{7}$, 49, $\frac{1}{343}$?

30a. To what base does $\sqrt{8}$ have logarithms $\frac{3}{4}$, -3, -1, $\frac{2}{3}$?

30b. To what base does $\sqrt[3]{25}$ have logarithms $\frac{2}{3}$, $-\frac{3}{2}$, -1, -2?

31. What is $\log_a 1$?

32a. What is $\lg_{10}\infty$?

32b. What is $\lg_{0.1}\infty$?

33a. What is $\lg_{10} 0$?

33b. What is $\lg_{0.1} 0$?

34. Which is greater: $\log_{14} 15$ or $\log_{15} 14$?

If some number is composed of other numbers by multiplying, dividing, raising to a power, or extracting a root, then the logarithm of such a number can be computed from the logarithms of the numbers of which it is composed; for example, the logarithms of the numbers ab and $\frac{a}{b}$ can be computed from the logarithms of a and b. This process is called *taking the logarithm* of the given quantity. The operation is performed on the basis of the following four theorems:

$$1)\ \log_a(MN) = \log_a M + \log_a N; \qquad 2)\ \log_a\left(\frac{M}{N}\right) = \log_a M - \log_a N;$$

$$3)\ \log_a(N^k) = k\log_a N; \qquad 4)\ \log_a \sqrt[k]{N} = \frac{1}{k}\log_a N,$$

i.e.: 1) the logarithm of a product equals the sum of the logarithms of the factors; 2) the logarithm of a fraction equals the difference between the logarithm of the numerator and the logarithm of the denominator; 3) the logarithm of a power equals the exponent multiplied by the logarithm of the base of the power; 4) the logarithm of a root equals the logarithm of the radicand divided by the index of the root.

35a. Express $\log 6$ through $\log 2$ and $\log 3$.

35b. Express $\log 21$ through $\log 3$ and $\log 7$.

36a. Express $\log 1\frac{2}{3}$ through $\log 5$ and $\log 3$.

36b. Express $\log 2\frac{3}{5}$ through $\log 13$ and $\log 5$.

37a. Express $\log 125$ through $\log 5$.

37b. Express $\log 81$ through $\log 3$.

38a. Express $\log \sqrt[4]{11}$ through $\log 11$.

38b. Express $\log \sqrt[5]{2}$ through $\log 2$.

39a. If the base of logarithms equals 3, then $\log 81 = 4$ and $\log 243 = 5$. What are $\log(81 \cdot 243)$ and $\log\frac{81}{243}$ to the same base?

39b. If the base of logarithms equals 2, then $\log 64 = 6$ and $\log 1024 = 10$. What are $\log(1024 \cdot 64)$ and $\log\frac{64}{1024}$ to the same base?

40a. The logarithms of which initial numbers must one know in order to find, to the same base, the logarithms of the numbers 24, $\frac{125}{27}$, $\sqrt{38}$, $\sqrt[3]{\frac{7}{25}}$?

40b. The same for the numbers 18, 27, 36, 40, 50.

41a. Knowing that $\lg 2 = 0.30103$, $\lg 3 = 0.47712$, and $\lg 5 = 0.69897$, find $\lg 6$, $\lg 15$, $\lg 30$, $\lg 10$, $\lg 1000$.

41b. Knowing that $\lg 2 = 0.30103$, $\lg 5 = 0.69897$, and $\lg 7 = 0.84510$, find $\lg 14$, $\lg 35$, $\lg 50$, $\lg 100$, $\lg 10000$.

42a. With the data of the preceding problem find $\lg 2\frac{1}{2}$, $\lg 1\frac{2}{3}$, $\lg \frac{2}{25}$, $\lg 0.6$, $\lg 0.016$.

42b. With the same data find $\lg 2\frac{4}{5}$, $\lg \frac{2}{7}$, $\lg \frac{5}{14}$, $\lg 0.07$, $\lg 0.0014$.

43a. With the data of problem 41 find $\lg 20$, $\lg 200$, and also $\lg 15$, $\lg 150$, $\lg 1500$.

43b. With the data of problem 41 find $\lg 70$, $\lg 700$, and also $\lg 35$, $\lg 350$, $\lg 3500$.

44a. With the data of problem 41 find $\lg 0.3$, $\lg 0.003$, $\lg 0.06$, $\lg 0.0006$.

44b. With the data of problem 41 find $\lg 0.2$, $\lg 0.002$, $\lg 0.14$, $\lg 0.0014$.

Take logarithms of the following expressions:

45a. $x = 2ab$.

45b. $x = 3bc$.

46a. $x = \frac{ab}{c}$.

46b. $x = \frac{a}{bc}$.

47a. $x = a^3b^2$.

47b. $x = a^2bc^3$.

48a. $x = \frac{a^2}{b^3c^7}$.

48b. $x = \frac{a^5b^6}{c^4}$.

49a. $x = 2(a+b)$.

49b. $x = 5(a-b)$.

50a. $x = \frac{3}{a^2-b^2}$.

50b. $x = \frac{a^2-b^2}{7}$.

51a. $x = \frac{(a-b)^2c}{(a+b)d}$.

51b. $x = \frac{a(b+c)}{(b-c)^2d}$.

52a. $x = 5a^2b\sqrt[3]{c}$.

52b. $x = 2b\sqrt{ac}$.

53a. $x = \sqrt{\frac{3a^3b}{c^4}}$.

53b. $x = \sqrt[4]{\frac{a^3}{2b^2c}}$.

54a. $x = 5a\sqrt[3]{a^2(a-b)}$.

54b. $x = 8a^3\sqrt[5]{a(b+c)^2}$.

55a. $x = \frac{2ab^3}{c\sqrt{d}}$.

55b. $x = \frac{a^2\sqrt[3]{b}}{c\sqrt{d}}$.

56a. $x = \left(\sqrt[3]{\frac{10}{a}}\right)^5$.

56b. $x = \left(\frac{\sqrt[3]{a}}{\sqrt{1000}}\right)^7$.

57a. $x = \frac{1}{a^n\sqrt{b}}$.

57b. $x = \frac{1}{a\sqrt[n]{b}}$.

58a. $x = \frac{1}{a}\sqrt{\frac{1}{b}}$.

58b. $x = \sqrt{\frac{1}{a}} \cdot \sqrt[3]{\frac{1}{b}}$.

59a. $x = a^{\frac{2}{3}}b^{\frac{3}{5}}$.

59b. $x = a^{-2}b^{\frac{4}{3}}$.

60a. $x = \left(\frac{m}{n}\right)^{\frac{m}{n}}$.

60b. $x = \left(\frac{1}{a}\right)^{-\frac{p}{q}}$.

61a. $x = \sqrt{2\sqrt{6\sqrt{15}}}$.

61b. $x = \sqrt[3]{3\sqrt[3]{21\sqrt[3]{6}}}$.

62a. $x = \sqrt[3]{\frac{a^2 b}{\sqrt[5]{c^3}}}.$

62b. $x = \sqrt[5]{\frac{a^3 \sqrt[3]{b}}{b^2}}.$

63a. $x = \frac{a^{-\frac{3}{4}} b^2}{c^{-\frac{1}{5}}}.$

63b. $x = \frac{a^{\frac{2}{5}} b^{-2}}{c^{-\frac{3}{4}}}.$

64a. $x = \sqrt{\frac{24\sqrt{2\sqrt{3}}}{\sqrt[3]{4\sqrt{6}}}}.$

64b. $x = \sqrt{\frac{15\sqrt{3\sqrt{5}}}{\sqrt{25\sqrt{3}}}}.$

65a. $x = \sqrt[3]{\frac{a}{\sqrt{ab}}} \cdot \sqrt[-2]{\frac{a}{b}}.$

65b. $x = \sqrt{\frac{\sqrt[5]{ab}}{b}} \cdot \sqrt[-2]{\frac{b}{a}}.$

66a. $x = \sqrt[m]{a\sqrt[n]{b}}.$

66b. $x = \sqrt[n]{a\sqrt[n]{b\sqrt[n]{c}}}.$

67a. $x = \sqrt[-m]{a\sqrt[-n]{b}}.$

67b. $x = \sqrt[-n]{a\sqrt[-n]{b\sqrt[-n]{c}}}.$

68a. $x = \left(\sqrt{2}\right)^{\sqrt{2}}.$

68b. $x = 0.5^{\sqrt[5]{0.5}}.$

69a. $x = \log\left(\sqrt{2}^{\sqrt{2}}\right).$

69b. $x = \log\left(0.5^{\sqrt[5]{0.5}}\right).$

70a. $x = \log\left(10^{ab}\right).$

70b. $x = \log\left(a^{2n}\right).$

71a. $x = \frac{10\log a}{\log(a^3)}.$

71b. $x = \frac{\log(a^2)}{\log^2 a}.$

72a. $x = \log\left[(a+b)^{\log(a+b)}\right].$

72b. $x = \log\sqrt{\log(b)^a}.$

73a. $x = \log\left(\sqrt[5]{a^4}^{\sqrt[3]{a^2}}\right).$

73b. $x = \log\left(\sqrt[3]{a^5}^{\sqrt[7]{a^1}}\right).$

74a. $x = \log\frac{\sqrt{(a+b)^{2\log(a-b)}}}{\sqrt{(a-b)^{\log(a+b)}}}.$

74b. $x = \log\frac{\sqrt[3]{(a^2+b^2)^{5\log(a-b)}}}{\sqrt[3]{(a-b)^{\log(a^2+b^2)}}}.$

By the logarithm of a number one can find the original number by operations inverse to the four rules of taking logarithms. The operation inverse to taking logarithms is called *potentiation.*

Find the number from their logarithms:

75a. $\log x = \log 7 - \log 3 + \log 2.$

75b. $\log x = \log 3 + \log 5 - \log 2.$

76a. $\log x = 3\log 5 + 2\log 3.$

76b. $\log x = 2\log 3 + 5\log 2.$

77a. $\log x = \frac{3}{5}\log 11 - \frac{2}{7}\log 5.$

77b. $\log x = \frac{1}{3}\log 17 - \frac{5}{9}\log 3.$

78a.
$\log x = 2\log 13 - \frac{2}{5}\log 2 - \frac{4}{3}\log 7.$

78b.
$\log x = 3\log 5 - \frac{7}{3}\log 19 - \frac{2}{3}\log 2.$

79a.
$\log x = 3\log a + 2\log b - 4\log c.$

79b. $\log x = \log a - 3\log b + \log c.$

80a.
$\log x = \frac{1}{2}\log a + \frac{1}{3}\log b - \frac{1}{4}\log c.$

80b. $\log x = \frac{m}{n}\log a + \frac{p}{q}\log b.$

81a. $\log x = \frac{\log 10}{2} - \frac{\log 5}{3}.$

81b. $\log x = \frac{\log a}{m} - \frac{\log b}{n}.$

82a.
$\log x = \frac{2}{5}\log(a+b) - \frac{3}{4}\log(a-b).$

82b.
$\log x = \frac{3}{2}\log(a-b) - \frac{5}{3}\log(a+b).$

83a. $\log x = \log(a+b) - \frac{2}{3}\left(2\log a + \frac{3}{4}\log b\right).$

83b. $\log x = 2\log(a-b) + \frac{3}{4}\left(\log a - \frac{2}{3}\log b\right).$

84a. $\log x = -3\log a + \frac{1}{3}\left[\log(a+b) + \frac{2}{5}\log(a-b) - \log b - \frac{1}{2}\log c\right].$

84b. $\log x = -\frac{2}{3}\log b + \frac{3}{4}\left[\log a - 2\log c - \log(a-b) + \frac{3}{5}\log(a+b)\right].$

Decimal Logarithms

85a. Knowing that $\lg 2 = 0.30103$, find the logarithms of the numbers 20, 2000, 0.2, and 0.000002.

85b. Knowing that $\lg 3 = 0.47712$, find the logarithms of the numbers 300, 3000, 0.03, and 0.0003.

86a. Knowing that $\lg 5 = 0.69897$, find the logarithms of the numbers 2.5; 500; 0.25 and 0.005.

86b. Knowing that $\lg 7 = 0.84510$, find the logarithms of the numbers 0.7, 4.9, 0.049, and 0.0007.

87a. Knowing that $\lg 3 = 0.47712$ and $\lg 7 = 0.84510$, find the logarithms of the numbers 210, 0.021.

87b. Knowing that $\lg 2 = 0.30103$ and $\lg 7 = 0.84510$, find the logarithms of the numbers 140; 0.14.

88a. Knowing that $\lg 3 = 0.47712$ and $\lg 5 = 0.69897$, find the logarithms of the numbers 1.5, 0.12, and 0.36.

88b. Knowing that $\lg 5 = 0.69897$ and $\lg 7 = 0.84510$, find the logarithms of the numbers 3.5, $\frac{5}{7}$, 0.28, $\frac{5}{49}$, and 1.96.

89a. Find the logarithms of the numbers 8, 141, 954, 420, 640, 1235, 3907, 3010, 18.43, 2.05, 900.1, 0.73, 0.0028, 0.1008, 0.00005.

89b. Find the logarithms of the numbers 15, 154, 837, 510, 5002, 1309, 8900, 8.315, 790.7, 0.09, 0.6745, 0.000745, 0.04257, 0.00071.

90a. Find the logarithms of the numbers 2174.6, 1445.7, 2169.5, 8437.2, 46.472, 6.2853, 0.78938, 0.054294, 631.074, 2.79556, 0.747428, 0.00237158.

90b. Find the logarithms of the numbers 2578.4, 1323.6, 8170.5, 6245.3, 437.65, 87.268, 0.059372, 0.84938, 62.5475, 131.037, 0.593946, 0.00234261.

91a. Find the numbers corresponding to the logarithms 3.16227, 3.59207, 2.93318, 0.41078, 1.60065, $\bar{2}.75686$, $\bar{3}.23528$, $\bar{1}.79692$, $\bar{4}.87806$, $\bar{5}.14613$.

91b. Find the numbers corresponding to the logarithms 3.07372, 3.69205, 1.64904, 2.16107, 0.70364, 1.31952, $\bar{4}.30814$, 3.00087, $\bar{2}.69949$, $\bar{6}.57978$.

92a. Find the numbers corresponding to the logarithms 3.57686, 3.16340, 2.40359, 1.09817, 4.49823, $\bar{2}.83882$, $\bar{1}.50060$, $\bar{3}.30056$, $\bar{1}.17112$, $\bar{4}.25100$.

92b. Find the numbers corresponding to the logarithms 3.33720, 3.09875, 0.70093, 4.04640, $\bar{1}.41509$, $\bar{2}.32649$, $\bar{4}.14631$, $\bar{3}.01290$, $\bar{5}.39003$.

93a. Convert the following logarithms into artificial form: -2.69537, -4.21293, -0.54225, -1.68307, -3.53820, -5.89990.

93b. Convert the following logarithms into artificial form: -3.21729, -1.73273, -5.42936, -0.51395, -2.43780, -4.22990.

94a. Find the true values of the logarithms $\overline{1}.33278$, $\overline{3}.52793$, $\overline{2}.95426$, $\overline{4}.23725$, $\overline{1}.39420$, $\overline{5}.67990$.

94b. Find the true values of the logarithms $\overline{2}.45438$, $\overline{1}.73977$, $\overline{3}.91243$, $\overline{5}.12912$, $\overline{2}.83770$, $\overline{4}.28990$.

Compute the following expressions by means of logarithms:

95a. $311 \cdot 25.6$.

95b. $4.51 \cdot 215$.

96a. $758 \cdot 0.53$.

96b. $0.037 \cdot 269$.

97a. $6603 : 213$.

97b. $8132 : 338$.

98a. $3.264 : 0.078$.

98b. $23.65 : 0.94$.

99a. 23.5^2.

99b. 11.8^2.

100a. 0.028^3.

100b. 0.0067^3.

101a. $\sqrt{12.5}$.

101b. $\sqrt{23.2}$.

102a. $\sqrt[3]{0.052}$.

102b. $\sqrt[3]{0.61}$.

103a. $\frac{483.6 \cdot 2.138}{25.58}$.

103b. $\frac{47.54 \cdot 3.642}{145.4}$.

104a. $\frac{0.045 \cdot 7.513}{2.071 \cdot 0.864}$.

104b. $\frac{14.5 \cdot 0.0178}{0.83 \cdot 3.105}$.

105a. $\sqrt[10]{34.567}$.

105b. $\sqrt[7]{71.238^3}$.

106a. $\sqrt[9]{0.06432}$.

106b. $\sqrt[8]{0.75^{15}}$.

107a. $5\sqrt[11]{3.1866}$.

107b. $2\sqrt[13]{2.7892}$.

108a. $\frac{109}{716}\sqrt{\frac{76}{93}}$.

108b. $\frac{21}{37}\sqrt{\frac{119}{295}}$.

109a. 1.04^{100}.

109b. 2.08^{30}.

110a. $\sqrt[100]{100}$.

110b. $\sqrt[200]{50}$.

111a. $\sqrt[7]{0.098756^3}$.

111b. $\sqrt[5]{0.98437^2}$.

112a. $\sqrt[3]{\left(\frac{37}{2939}\right)^5}$.

112b. $\sqrt{\left(\frac{43}{7243}\right)^4}$.

113a. $\left(8.53\sqrt[10]{10}\right)^{\frac{2}{3}}$.

113b. $\left(2.38\sqrt[5]{10}\right)^{\frac{3}{5}}$.

114a. $\left(\frac{38}{27}\right)^{0.07}\left(\frac{51}{43}\right)^{0.03}$.

114b. $\left(\frac{25}{7}\right)^{0.03}\left(\frac{39}{19}\right)^{0.07}$.

115a. $\sqrt{0.006\sqrt{0.17624}}$.

115b. $\sqrt{0.89394\sqrt{0.092}}$.

116a. $\sqrt[5]{0.4293\sqrt{\frac{19}{34}}}$.

116b. $\sqrt[8]{\frac{37}{43}\sqrt[5]{0.3798}}$.

117a. $\frac{1}{0.7345^3 \cdot 0.164^2}$.

117b. $\frac{1}{0.2127^2 \cdot 0.921^3}$.

118a. $\sqrt[3]{0.054\sqrt[3]{0.0003617}}$.

118b. $\sqrt[5]{0.0007\sqrt{0.09342}}$.

119a. $\left(\frac{4}{7}\right)^{0.45}$.

119b. $\sqrt[-3]{1\frac{1}{3}}$.

120a. $\sqrt[15]{\left(\frac{1}{8}\right)^{1.1}}$.

120b. $\sqrt[0.8]{2.39977^{-9}}$.

121a. $\frac{0.0045 \cdot 7.5132}{2.0719 \cdot 0.864}$. **121b.** $\frac{14.51 \cdot 0.017085}{0.78 \cdot 3.1057}$.

122a. $\frac{3.5216^3 \cdot 0.027^2}{0.21785}$. **122b.** $\frac{40.12^2 \cdot 0.0113^3}{0.98763}$.

123a. $\sqrt[9]{\frac{8}{7}\sqrt[6]{54321}}$. **123b.** $\sqrt{\frac{7}{5}\sqrt[4]{23468}}$.

124a. $\frac{0.0875}{9.8304}\sqrt{\frac{78}{0.007615}}$. **124b.** $\frac{0.0379}{2.4548}\sqrt{\frac{123}{0.009843}}$.

125a. $\sqrt{\frac{2\sqrt[10]{118098}}{3\sqrt{10}}}$. **125b.** $\sqrt[3]{0.261255\sqrt{\frac{2}{7}}}$.

126a. $\frac{0.170586^4 \cdot \sqrt[5]{0.2}}{0.06561^3 \cdot \sqrt[6]{0.1}}$. **126b.** $\frac{0.9^{-6} \cdot 5120^{0.4}}{\sqrt[3]{229376}}$.

127a. $\frac{5^{10}\sqrt{0.01059137}}{12.631}$. **127b.** $\frac{10.59952^{10} \cdot \sqrt[3]{0.9}}{67851.3}$.

128a. $\frac{8.36\sqrt{0.0067254}}{0.96578\sqrt[3]{0.000035746}}$. **128b.** $\frac{2.79\sqrt[3]{0.0029745}}{0.79438\sqrt{0.000054237}}$.

129a. $\frac{87.285^2 \cdot \sqrt[10]{75.846}}{\sqrt[3]{-3.055}}$. **129b.** $\frac{29.348^2 \cdot \sqrt{93.594}}{\sqrt[5]{-2.743}}$.

130a. $\sqrt[5]{\frac{0.03425\sqrt[7]{136}}{0.00034}}$. **130b.** $\sqrt[4]{\frac{0.26758\sqrt[3]{0.4}}{0.006422}}$.

131a. $\sqrt[10]{58} \cdot 0.3599545^{-5}$. **131b.** $\sqrt[8]{0.06084925} \cdot 0.63^{-10}$.

132a. $\frac{\left(-\sqrt[5]{322625}\right)^6}{\left(\sqrt[7]{10732574}\right)^{-4}}$. **132b.** $\frac{1}{\left(0.03029067 \cdot \sqrt{4834.719}\right)^{\frac{3}{5}}}$.

133a. $\frac{109}{716}\sqrt[3]{\frac{95\pi}{404}}$. **133b.** $\frac{21}{37}\sqrt[4]{\frac{71\pi}{1503}}$.

134a. $\left(9.8\pi\sqrt[5]{1.6325}\right)^{\frac{2}{3}}$. **134b.** $\left(1.6224\sqrt[3]{4\pi}\right)^{\frac{3}{5}}$.

135a. $0.0009^{0.0009}$. **135b.** $0.0007^{0.0007}$.

136a. $0.0376^{0.0376}$. **136b.** $0.0289^{0.0289}$.

137a. $\sqrt[7.032]{0.4275}$. **137b.** $\sqrt[3.271]{0.2837}$.

138a. $0.513^{\sqrt[5]{0.69537}}$. **138b.** $0.29342^{\sqrt[7]{0.4126}}$.

139. The area of a right triangle is 282.14 cm^2; one of its legs is three times the other. Find the legs.

140. The volume of a rectangular parallelepiped is 385 m^3. The dimensions of the parallelepiped are in the ratio 3 : 5 : 8. Find the dimensions.

141. The volume of a regular triangular pyramid is 187.2 cm^3. The side of its base is half its height. Determine the side of the base.

142. Find the edge of a cube having a volume twice the volume of a cube with edge 2.378 m.

143. Find the volume of a sphere whose surface equals 0.038 m^2.

144. Find the radius R of the base of a cone having volume $v = 36.785$ dm^3 and height $H = R$.

145. Find the edge of a cube having volume $v = 0.5$ m^3.

146. A wooden sphere, floating in water, displaced 2791.22 cm^3 of water. Determine the radius of the sphere (specific gravity of wood 0.5).

147. It is required to cast a lead sphere of volume 2.332 dm^3. Determine the diameter of the casting (mold).

148. The volume of a cone is 96 cm^3. The radius of its base is one quarter of its height. Determine the radius of the base of the cone.

149. Find the degree measure of an arc of a circle of length $S = 115.3$ m at a circle radius $R = 667.5$ m.

150. How much will a bronze telephone wire of length 1.25 km weigh, if its thickness is 2.7 mm and the specific gravity of bronze is 9?

151. The mean density of the terrestrial globe is 5.54 g. Find the mass of the earth, if its mean radius equals 6371 km.

152. Using the formula $a = \frac{4\pi^2 R}{T^2}$, find the acceleration a of the centrifugal force of the Moon, if its distance from the Earth is $R = 60.27$ earth radii and the time of revolution around the Earth is $T = 27.32$ days.

153. Using the formula of uniformly accelerated motion $s = \frac{gt^2}{2}$, find the path s travelled by a body during 11 seconds, if the acceleration $g = 9.8$.

154. The quantity of heat in small calories liberated in t seconds by a current of i amperes with a conductor resistance of r ohms is expressed by the formula $Q = 0.24i^2rt$. How much heat is liberated in 1 hour by a 16-candle bulb consuming a current of 0.315 ampere, if the resistance of the bulb's filament equals 357.2 ohms?

155. Find the edge of a regular tetrahedron whose volume is $v = 1$ dm^3.

156. Find the density of steel, if from experiment it was found that a steel cylinder with dimensions $R = 5.5$ cm and $H = 16.5$ cm weighs 13.09 kg.

157. A coil of steel wire weighs 20 kg. The diameter of the wire is 2.5 mm. Determine the length of this wire (taking the specific gravity of steel for wire as 7.96).

158. How many liters of gas does a balloon hold, having the shape of a sphere whose radius is $R = 1.738$ dm? *Hint.* $\lg \frac{4}{3}\pi$ is given in the table "Some constant numbers."

159. The time of one swing of a pendulum is determined by the formula $t = \pi\sqrt{\frac{l}{g}}$, where l is the length of the pendulum and g the acceleration of a freely falling body. For Moscow $g = 9.815$ m/sec^2. Compute for Moscow the length of the seconds pendulum (i.e. a pendulum making one swing in one second).

If an expression contains operations of addition and subtraction, it can be logarithmed only after being reduced to a form convenient for taking logarithms. If reducing the expression to such a form presents great difficulty, one must separate the individual parts connected by signs of addition or subtraction, compute these parts by taking logarithms, substitute the obtained results into the initial expression, add and subtract these results (without taking logarithms, of course), and finally compute the new form of the initial expression (also mostly by taking logarithms).

160a. $\sqrt{145.27^2 - 124.49^2}$. **160b.** $\sqrt[3]{273.43^2 - 111.21^2}$.

161a. $\sqrt[6]{8 - \sqrt[5]{10}}$. **161b.** $\sqrt[5]{21 - \sqrt[3]{17}}$.

162a. $\sqrt{11.367 - \sqrt[3]{16.729}}.$ **162b.** $\sqrt[3]{53.114 - \sqrt{15.277}}.$

163a. $\sqrt[10]{2.1663 - \sqrt[11]{4919.6}}.$ **163b.** $\sqrt[7]{1.5947 - \sqrt[10]{237.53}}.$

164a. $\frac{1}{0.239^3+0.083^5}.$ **164b.** $\frac{1}{0.0375^2+0.597^3}.$

165a. $\sqrt[16]{\frac{43+5\sqrt[3]{268}}{\sqrt{17}}}.$ **165b.** $\sqrt[11]{\frac{12+7\sqrt[5]{277}}{\sqrt[3]{11}}}.$

166. $\sqrt{\frac{17569}{111.11}} - \sqrt[3]{\frac{67685}{1.2365}}.$

167. $\sqrt[20]{\frac{27+3\sqrt[20]{1.4762}}{\sqrt[5]{11}}}.$

168. $\sqrt{0.859^3 + 5\sqrt[3]{11}}.$

169. $\sqrt[13]{2.459^{6.3} + 8.742^{2.3}}.$

170a. $\sqrt[7]{\frac{\sqrt{2}-\sqrt[3]{11}}{3^{0.561}}}.$ **170b.** $\sqrt[5]{\frac{\sqrt[3]{3}-\sqrt[5]{21}}{5^{0.692}}}.$

171a. $\sqrt[-3.2]{(6.263 + \sqrt[3]{-4.94623})^5}.$ **171b.** $\sqrt[-2.3]{(2.798 + \sqrt{31.5946})^3}.$

172a. $\sqrt[9]{(\sqrt[4]{0.723} + \sqrt[1.6]{1.23794})^{-2}}.$ **172b.** $\sqrt[6]{(\sqrt[5]{0.989} + \sqrt[1.8]{2.54932})^{-5}}.$

173a. $\frac{\sqrt[5]{0.8\sqrt[3]{0.7-(1.2686)^{-2}}}}{\sqrt[20]{0.0874968^3}}.$ **173b.** $\frac{\sqrt[3]{0.3\sqrt[5]{0.11-(1.6967)^{-4}}}}{\sqrt[10]{0.374932^3}}.$

174a. $\frac{\sqrt[-4]{1.2-(1.2368)^{-0.72}}}{(\sqrt[5]{0.423286-0.87})^2}.$ **174b.** $\frac{\sqrt[-5]{2.37-(3.2143)^{-0.67}}}{(\sqrt{0.597296-0.713})^3}.$

Exponential and Logarithmic Equations

An equation is called *exponential* if the unknown enters into the exponent of a power or root. An equation is called *logarithmic* if the unknown enters under the sign of a logarithm. The principal methods of solving these equations are: 1) equalizing the bases; 2) taking logarithms of both sides of the equation, without using or with the use of logarithm tables; 3) potentiating both sides of the equation; 4) reducing the exponential or logarithmic equation to an algebraic one with respect to an expression containing the exponential or logarithmic function of the unknown. For example:

I. $2^x = 8;\ 2^x = 2^3;\ x = 3.$

II. $x^{\lg x} = 10;\ \lg^2 x = 1;\ \lg x = \pm 1;\ x = 10$ or $0.1.$ $2^x = 9;\ x\lg 2 = \lg 9;$ $x \cdot 0.30103 = 0.95424;\ x = \frac{95424}{30103} \approx 3.17.$

III. $2\lg x - 1 = 0;\ \frac{x^2}{10} = 1;\ x^2 = 10;\ x = \sqrt{10}$ (here one cannot take the radical with the minus sign, since with a decimal base a negative number has no logarithm).

IV. $10 \cdot 2^x - 2^{2x} = 16;\ 2^{2x} - 10 \cdot 2^x + 16 = 0;\ 2^x = 5 \pm \sqrt{25 - 16} = 8$ and $2;$ $2^{x_1} = 8$, whence $x_1 = 3;$ $2^{x_2} = 2$, whence $x_2 = 1.$

175. $10^{-x} = 10\,000.$

176. $\sqrt[3]{a^x} = \sqrt{a^{3x+2}}.$

177. $16^x = \frac{1}{4}.$

178. $\sqrt[1-x]{a^3} = \sqrt[3-x]{a^2}.$

179. $\left(\frac{4}{9}\right)^x = \left(\frac{3}{2}\right)^{-5}$.

180. $\sqrt{a^{x-1}}\,\sqrt[3]{a^{2x-1}}\,\sqrt[4]{a^{2-8x}} = 1$.

181. $\left(\frac{1}{0.125}\right)^x = 128$.

182. $a^{(1-x)(x-2)} = \frac{1}{a^6}$.

183. $\sqrt[x]{256} = 4^x$.

184. $2^x - 2^{x-2} = 3$.

185. $2^{2x} \cdot 3^x = 144$.

186. $5^{x+1} + 5^x = 750$.

187. $10^{(3-x)(4-x)} = 100$.

188. $\sqrt[a-x]{c^{b+x}} = \sqrt[a+x]{c^{b-x}}\,\sqrt[a^2-x^2]{c^2}$.

189. $4^{\sqrt{x+1}} = 64 \cdot 2^{\sqrt{x+1}}$.

190. $5^{(x^2+x-2)(3-x)} = 1$.

191. $2^{x^2-7.7x+16.5} = 8\sqrt{2}$.

192. $\sqrt{9^{x(x-1)-\frac{1}{2}}} = \sqrt[4]{3}$.

193. $92^{\lg x} = 778\,688$.

194. $\sqrt[x-1]{4096} = 2\sqrt[x]{32\,768}$.

195. $5 \cdot \sqrt[x+2]{3125^{x+1}} = \sqrt[x+3]{15625^{x+2}}$.

196. $6^{2x+4} = 3^{3x} \cdot 2^{x+8}$.

197. $6^x + 6^{x+1} = 2^x + 2^{x+1} + 2^{x+2}$.

198. $5^{1-x} = 7^{x-1}$.

199. $\left(\frac{3}{4}\right)^{x-1} \sqrt{\frac{4}{3}} = \frac{1}{2}\left(\sqrt[4]{3}\right)^{3x-4}$.

200. $5^{2x-1} + 2^{2x} - 5^{2x} + 2^{2x+2} = 0$.

201. $x^x = x$.

202. $x^{\sqrt{x}} = (\sqrt{x})^x$.

203. $3 \cdot 2^x = 4\sqrt[x]{9}$.

204. $x^{\lg x} = 10$.

205. $x^{\lg x} = 100x$.

206. $\sqrt[3]{x^{\lg x - 1}} = 100$.

207. $10^x = \sqrt[x]{5}$.

208. $5^x = 17$.

209. $10^x = 200$.

210. $\left(\frac{2}{3}\right)^x = 8$.

211. $23^x = 100$.

212. $10^x = \sqrt[x]{2}$.

213. $52^x = 0.1$.

214. $\sqrt[x]{1.3713} = \sqrt[10]{10}$.

215. $3^x - 5^{x+2} = 3^{x+4} - 5^{x+3}$.

216. $7^{x-1}+7^{x-2}+7^{x-3}=5^{x-1}+5^{x-2}+5^{x-3}$.

217. $\left(\frac{1}{8}\right)^x=5^{10}$.

218. $\sqrt{0.35^x}=0.00007882$.

219. $\lg x=1-\lg 3$.

220. $\lg x=\lg 24-\lg 8$.

221. $\lg x=3\lg 18-4\lg 12$.

222. $\frac{\lg x}{1-\lg 2}=2$.

223. $1-\lg 5=\frac{1}{3}\left(\lg\frac{1}{2}+\lg x+\frac{1}{3}\lg 5\right)$.

224. $\lg\left(2x-\frac{9}{4}\right)-\lg x=\lg(x-3)$.

225. $\lg\left(x-\frac{8}{9}\right)=2\lg\frac{1}{6}$.

226. $\lg(x-2)=\lg x-\lg 2$.

227. $\lg\left(3x^2+7\right)-\lg(3x-2)=1$.

228. $\lg\left(x+\sqrt{3}\right)=-\lg\left(x-\sqrt{3}\right)$.

229. $\frac{\lg x}{\lg(x+1)}=-1$.

230. $\frac{1}{2}\lg(x-9)+\lg\sqrt{2x-1}=1$.

231. $\lg 10+\frac{1}{3}\lg\left(271+3^{\sqrt{2x}}\right)=2$.

232. $\log_a\log_a x=\log_a m+\log_a n$.

233. $\log_a\log_a x=\log_a\log_a m-\log_a n$.

234. $a^{2x}+c^2=2ba^x$.

235. $5^{2x}-5^x=600$.

236. $3^{2x+5}=3^{x+2}+2$.

237. $0.1\lg^4 x+0.9=\lg^2 x$.

238. $8^{x+1}-8^{2x-1}=30$.

239. $\begin{cases}5x+2y=100;\\ \lg x-\lg y=\lg 1.6.\end{cases}$

240. $\begin{cases}\lg x+\lg y=7;\\ \lg x-\lg y=5.\end{cases}$

241. $\begin{cases}14^x=63y;\\ 17^x=87y.\end{cases}$

242. $\begin{cases}x^y=y^x;\\ x^2=y^3.\end{cases}$

243. $\begin{cases}x^{x+y}=y^{12};\\ y^{x+y}=x^3.\end{cases}$

244. $\begin{cases}0.4^{x+y}=\left(\frac{2}{5}\right)^2;\\ 1.4^{x-y}=1.6565.\end{cases}$

245. $\begin{cases} x^{\sqrt{y}} = y; \\ y^{\sqrt{y}} = x^4. \end{cases}$

246. $\begin{cases} x^{\sqrt{x}-\sqrt{y}} = y^4; \\ y^{\sqrt{x}-\sqrt{y}} = x. \end{cases}$

247. $\begin{cases} x^y = 243; \\ \sqrt[y]{1024} = \left(\frac{2}{3}x\right)^2. \end{cases}$

248. $\begin{cases} 3^y \sqrt[x]{64} = 36; \\ 5^y \sqrt[x]{512} = 200. \end{cases}$

Compound Interest Problems

249a. Into what sum will a deposit of 246 rubles, placed in a bank for 8 years at 5%, grow?

249b. Into what sum will a deposit of 3768 rubles, placed in a bank for 20 years at 4%, grow?

250a. How much must be deposited in a bank paying 6% per year in order to have 8000 rubles in 20 years?

250b. How much must be deposited in a bank paying 3% per year in order to have 6720 rubles in 12 years?

251a. In how many years will a capital of 20728 rubles grow into 50000 rubles, reckoning at $4\frac{1}{2}$%?

251b. In how many years will a deposit of 18978 rubles grow into 48593 rubles, reckoning at $7\frac{1}{2}$%?

252a. At what rate of interest will a deposit of 2498 rub. 60 kop. grow into 4000 rubles in 12 years?

252b. At what rate of interest will a deposit of 2465 rubles grow into 4015 rub. 30 kop. in 10 years?

253. At what rate of interest will a deposit double in 10 years?

254. A building cooperative received a loan of 7025 rub. 16 kop., in payment of which it must contribute 12000 rubles over 10 years. At what rate of interest was the loan issued to the cooperative, reckoning compound interest?

255. How many years will a building cooperative have to pay off a loan of 8760 rubles, for which, by calculation at $5\frac{1}{2}$% compound interest, the cooperative will have to pay out 25000 rubles?

256. What constant sum must a factory administration deposit in a bank at the beginning of each year to form a capital for building a workers' club, so that by the end of the tenth year this capital reaches 60000 rubles, given that the bank credits 5% compound interest?

257. In what period does a deposit lent out at $6\frac{1}{4}$% compound interest grow fourfold?

258. What sum will accumulate from annual contributions of 600 rubles at the beginning of each year over 20 years at $4\frac{1}{2}$% compound interest?

259. Determine the size of the loan received by a cooperative, if to pay it off it must contribute 1500 rubles at the end of each year for 12 years, reckoned at

$7\frac{1}{2}\%$.

260. A building cooperative, to pay off a loan, collects from its members 8 rubles per month for each thousand rubles of their debt, by the calculation of paying off the loan over 15 years by annual contributions at the end of each year at 5%. Did the cooperative correctly apportion the payment of contributions among its members?

261. In how many years can a 6-percent loan of 14752 rubles be paid off by annual contributions of 3000 rubles at the end of each year?

262. A forest plot, according to estimate, contains 4850 m^3 of timber. By how much will the plot increase in 12 years, if the average annual growth of timber is 1.2%?

263. 14 years ago a forest plot contained 24500 m^3. During this time it grew to 33450 m^3. Determine the average annual percentage of growth.

264. A forest plot is worth 22000 rubles, and its value grows annually by $2\frac{1}{2}\%$. How much will it be worth in 15 years?

265. Certain pathogenic bacteria divide every half hour. How many bacteria could there be in an organism (assuming unimpeded growth) 24 hours after an infection in which 10 of them were introduced?

266. Moscow numbered 2.8 million inhabitants in 1931. It is supposed that by 1940 the population of Moscow will grow to 6 million people. What is the percentage of annual population growth?

267. The annual growth of a forest averages 2.25%. How much firewood can a forest plot yield in 80 years, if at present it contains 13490 m^3?

268. The annual growth of a forest is 3.5%. At present 14500 m^3 of firewood can be gathered in it. How many cubic meters of firewood will there be in it in 10 years?

269. A city at present has 120000 inhabitants, and 20 years ago it had 65000. In how many years will the city have 200000 inhabitants, if the percentage of population growth is taken as constant?

270. The gross output of factory industry of the USSR in 1926/27 prices was 15.7 billion rubles in 1928, and 34.3 billion rubles in 1932. Determine the average percentage of annual growth of output in the first five-year plan.

271. The urban population of the USSR was 27.6 million people in 1928, and in 1932 it increased to 38.7 million people. Determine the average annual growth in percent.

Chapter XVII

Combinatorics

Combinatorial selections are of three kinds: arrangements, permutations, and combinations.

Arrangements of a total number of elements taken several at a time are selections that differ from one another both in the elements themselves and in their order. The number of arrangements of n elements taken k at a time is denoted A_n^k, where

$$A_n^k = n(n-1)(n-2)\dots(n-k+1).$$

Permutations of n elements are selections of all n elements that differ from one another only in the order of the elements. The number of permutations of n elements is denoted P_n, where

$$P_n = A_n^n = n(n-1)(n-2)\dots 3\cdot 2\cdot 1 = 1\cdot 2\cdot 3\dots n.$$

Combinations of a total number of elements taken several at a time are selections that differ from one another only in the elements themselves and in which the order of the elements does not matter. The number of combinations of n elements taken k at a time is denoted C_n^k, where

$$C_n^k = \frac{A_n^k}{P_k} = \frac{n(n-1)(n-2)\dots(n-k+1)}{1\cdot 2\cdot 3\dots k}.$$

Besides these notations there are others: if one agrees to denote the product of the natural numbers from 1 to n by $n!$ (this product is called the *factorial*), then the formulas above take the form

$$A_n^k = \frac{n!}{(n-k)!}; \quad P_n = n!; \quad C_n^k = \frac{n!}{k!(n-k)!}.$$

1a. Form the permutations of three elements.

1b. Form the permutations of four elements.

2a. Form the arrangements of four elements taken three at a time.

2b. Form the arrangements of five elements taken three at a time.

3a. Form, by means of arrangements, the permutations of three elements.
4a. Form the arrangements of all kinds of four elements.
5a. Form the combinations of all kinds of four elements.
6a. Form, by means of combinations, the arrangements of all kinds of three elements.
7a. Express the arithmetic numbers A_7^3, P_5, C_6^4.
8a. Express the arithmetic numbers P_8, A_{13}^7, C_{21}^6.
9a. Express the number of arrangements of $n+1$ elements taken $k-1$ in each arrangement.
10a. Express the number of arrangements of $m+n$ elements taken $m-n+1$ in each arrangement.
11a. Verify the equalities $C_9^3 = C_9^6$ and $C_{12}^7 = C_{12}^5$.
12a. Verify the equalities $C_6^4 + C_6^3 = C_7^4$ and $C_{10}^6 + C_{10}^5 = C_{11}^6$.
13a. Express the number of combinations of $n+2$ elements taken $k-1$ in each combination.
14a. Express the number of combinations of $m-n$ elements taken $n+1$ in each combination.
15a. In how many ways can four people be seated at a table?
16a. In how many ways can four-color ribbons be formed from seven ribbons of different colors?
17a. In how many ways can four persons be chosen for four different posts from nine candidates for these posts?
18a. How many straight lines can be drawn between ten points positioned so that no three of them lie on one line?

3b. Form, by means of arrangements, the permutations of four elements.
4b. Form the arrangements of all kinds of five elements.
5b. Form the combinations of all kinds of five elements.
6b. Form, by means of combinations, the arrangements of all kinds of four elements.
7b. Express the arithmetic numbers A_8^5, P_6, C_{10}^7.
8b. Express the arithmetic numbers P_{11}, A_{15}^9, C_{18}^7.
9b. Express the number of arrangements of $n-2$ elements taken $k+1$ in each arrangement.
10b. Express the number of arrangements of $m-n$ elements taken $m-2n-1$ in each arrangement.
11b. Verify the equalities $C_8^5 = C_8^3$ and $C_{15}^7 = C_{15}^8$.
12b. Verify the equalities $C_7^5 + C_7^4 = C_8^5$ and $C_{12}^6 + C_{12}^5 = C_{13}^6$.
13b. Express the number of combinations of $n-1$ elements taken $k+2$ in each combination.
14b. Express the number of combinations of $m+n$ elements taken $n-2$ in each combination.
15b. In how many ways can five people be seated at a table?
16b. How many different three-digit numbers can be written by means of nine digits?
17b. In how many ways can four persons be chosen for four identical posts from nine candidates for these posts?
18b. How many circles can be drawn between ten points positioned so that no four of them lie on one circle?

19a. From how many objects can 210 arrangements of two objects each be formed?

19b. From how many objects can 66 different pairs be formed?

20a. How many objects must be taken so that the number of arrangements of them taken 4 at a time is 12 times the number of arrangements taken 2 at a time?

20b. How many objects must be taken so that the number of combinations of them taken 3 at a time is to the number of combinations taken 5 at a time as 2 : 3?

21a. The number of combinations of n elements taken 3 at a time is one fifth of the number of combinations of $n+2$ elements taken 4 at a time. Find n.

21b. The number of arrangements of n elements taken 5 at a time is 18 times the number of arrangements of $n-2$ elements taken 4 at a time. Find n.

22a. The number of combinations of $2n$ elements taken $n+1$ at a time is to the number of combinations of $2n+1$ elements taken $n-1$ at a time as 3 : 5. Find n.

22b. The number of combinations of $2n$ elements taken $n-1$ at a time is to the number of combinations of $2n-2$ elements taken n at a time as 77 : 20. Find n.

23a. Show that the direct determination of the number of pairwise combinations reduces to summing the differences of a progression.

23b. Show that the direct determination of the number of triple combinations reduces to summing a series of pairwise products.

24. Among the permutations of the digits of the number 12345, how many begin with the digit 1? with the number 12? with the number 123?

25. Among the combinations of the 10 letters a, b, c... taken 4 at a time, how many contain the letter a? the letters a and b?

26. Among the arrangements of the 12 letters a, b, c... taken 5 at a time, how many contain the letter a? the letters a and b?

27. Among the combinations of h letters taken k at a time, how many contain n specified letters each?

28. Among the arrangements of h letters taken k at a time, how many contain n specified letters each?

29. For which values of k, and for how many, does the inequality $C_n^{k-1} < C_n^k$ hold?

30. Show that for even n, in the series of numbers of combinations $C_n^1, C_n^2 \dots C_n^{n-1}$, there is one middle number, the largest of all.

Chapter XVIII

Newton's Binomial

Newton's binomial formula is

$$(x+a)^n = x^n + \tfrac{n}{1}ax^{n-1} + \tfrac{n(n-1)}{1\cdot 2}a^2x^{n-2} + \dots$$
$$+ \tfrac{n(n-1)}{1\cdot 2}a^{n-2}x^2 + \tfrac{n}{1}a^{n-1}x + a^n.$$

The expression for the general term of Newton's binomial: $T_{k+1} = C_n^k a^k x^{n-k}$.

Find, by the shortcut method, the products of the binomials:

1a. $(x+1)(x+2)(x+3)(x+4)$.

1b. $(x-1)(x-2)(x-4)(x-5)$.

2a. $(x-1)(x+3)(x-4)(x+5)$.

2b. $(x+2)(x-3)(x+4)(x-6)$.

3a. $(x+1)(x+2)(x+3)(x+4)(x+5)$.

3b. $(x-1)(x-2)(x-3)(x-4)(x-5)$.

4a. $(x-2)(x+3)(x-4)(x+5)(x-6)$.

4b. $(x+2)(x-3)(x-4)(x+5)(x-6)$.

Find the expansions of the powers of the binomials:

5a. $(a+b)^6$.

5b. $(a+b)^8$.

6a. $(a-b)^7$.

6b. $(a-b)^5$.

7a. $(a+1)^9$.

7b. $(a+1)^{12}$.

8a. $(1-a)^8$.

8b. $(1-a)^{11}$.

9a. $(a+b^2)^5$.

9b. $(a^2-b)^9$.

10a. $(a-2b)^8$.

10b. $(3b+a)^6$.

11a. $(\sqrt{a}+\sqrt{b})^6$.

11b. $(\sqrt[3]{a}-\sqrt[3]{b})^6$.

12a. $(\sqrt[3]{2a}-\sqrt[3]{3b})^5$.

12b. $(\sqrt{3a}+\sqrt{2b})^5$.

13a. Find the 5th term of the expansion of $(a-b)^9$.

13b. Find the 8th term of the expansion of $(a-b)^{15}$.

14a. Find the middle term of the expansion of $(a-b)^{14}$.

14b. Find the two middle terms of the expansion of $(a-b)^{17}$.

15a. In the expansion of $(x+a)^{19}$ find the terms that contain the letter a to the 8th power and the letter x to the 8th power.

15b. In the expansion of $(x+a)^{16}$ find the terms that contain the letter a to the 11th power and the letter x to the 11th power.

16a. In the expansion of $(x^2 - ax)^{24}$ find the terms whose coefficient is the number of combinations taken 18 at a time.

17a. In the expansion of $(\sqrt{z} + \sqrt[3]{z})^9$ find the term that, after simplification, contains the letter z to the 4th power.

18a. In the expansion of $\left(\frac{2z}{a^2} + \frac{a}{z}\right)^8$ find the term not containing z.

19a. Find the 5th term of the expansion of $(\sqrt{1+z} - \sqrt{1-z})^n$ if the coefficient of the 3rd term equals 78.

20a. The sum of the coefficients of the 2nd and 3rd terms of the expansion of $\left(\sqrt[3]{z^2} + z^{-1}\right)^n$ equals 78. Determine the term of the expansion not containing z.

16b. In the expansion of $(x^3 - a^2x)^{31}$ find the terms whose coefficient is the number of combinations taken 7 at a time.

17b. In the expansion of $\left(\sqrt[6]{z} + \sqrt[3]{z^2}\right)^{12}$ find the term that, after simplification, contains the letter z to the 6th power.

18b. In the expansion of $\left(\frac{z}{a} + \frac{3a^2}{z}\right)^{10}$ find the term not containing z.

19b. Find the 4th term of the expansion of $(\sqrt[3]{1+z} - \sqrt[3]{1-z})^n$ if the coefficient of the 3rd term equals 45.

20b. The sum of the coefficients of the 2nd and 3rd terms of the expansion of $\left(\sqrt[5]{z^2} + z^{-\frac{1}{6}}\right)^n$ equals 153. Determine the term of the expansion not containing z.

Chapter XIX

Divisibility of Polynomials

Without performing the division, determine whether it goes evenly or with a remainder:

1a. $(x^2 - 7x + 12) : (x - 3)$.

1b. $(x^2 - 7x + 12) : (x - 4)$.

2a. $(x^3 - x^2 + 4) : (x + 2)$.

2b. $(x^3 - 3x^2 + 6x) : (x + 3)$.

3a. $(x^3 - 2x^2 + 3x - 2) : (x - 1)$.

3b. $(3x^3 - 2x^2 + 7x - 3) : (x - 1)$.

4a. $(x^3 - 8) : (x - 2)$.

4b. $(x^3 - 125) : (x + 5)$.

5a. $(x^3 + 64) : (x + 4)$.

5b. $(x^3 + 27) : (x - 3)$.

6. $(x^4 + 16) : (x - 2)$.

7. $(x^5 - 243) : (x + 3)$.

Without performing the division, find the remainder of the division:

8a. $(2x^4 - 3x^3 - 7x + 6) : (x - 2)$.

8b. $(x^4 - 5x^2 + 6x - 1) : (x + 4)$.

9a.
$(2x^5 + 8x^4 - 4x^3 + 13) : (x + 1)$.

9b.
$(3x^4 - 6x^3 + 5x^2 - x - 1) : (x + 2)$.

Solve the equations:

10. $x^3 - x^2 + 6x = 0$.

11. $x^3 - 3x^2 + 6x - 4 = 0$.

12. $x^3 - 7x + 6 = 0$.

13. $x^3 + 2x^2 - 2x + 3 = 0$.

14. $x^4 + x^3 + 2x - 4 = 0$.

Factor the polynomials into linear factors:

15a. $x^3 - 4x^2 + x + 6$.

15b. $x^3 + 4x^2 + 2x - 8$.

16. $x^4 - 3x^3 - 8x^2 + 12x + 16$.

17. $x^4 - 3x^3 + x^2 + 3x - 2$.

Chapter XX

Inequalities

In the following examples, add the two given inequalities:

1. $5 > -3$, $8 > 5$.

2. $2 < 5$, $-7 < -3$.

3. $x^2 > a + 1$, $2x > a - 5$.

4. $3x + y < 2a + 1$, $3y - 2x < 14 - 2a$.

In the following examples, subtract the second inequality from the first:

5. $16 > 13$, $2 < 5$.

6. $-8 < -5$, $-2 > -7$.

7. $2x > b^2$, $a^2 < 9 - x$.

8. $(a - b)^2 < 2$, $(a + b)^2 > 8$.

Multiply both sides of the inequality by the indicated factors:

9. $5 > -2$ by 5.

10. $-7 < -5$ by -2.

11. $a^2 > b$ by $-b$.

12. $a - 1 < b$ by $-m$.

Divide both sides of the inequality by the indicated divisors:

13. $-6 < 9$ by 3.

14. $-15 > -35$ by -5.

15. $a^3 < a^2$ by $-a$.

16. $(a - b)^3 > (a - b)^2$ by $a - b$.

Multiply the inequalities term by term:

17. $5 > 3$, $7 > 2$.

18. $2 > -5$, $-3 > -7$.

19. $-3 < 5$, $-5 < 2$.

20. $-13 < -7$, $-9 > -15$.

Divide the inequalities:

21. $35 < 40$, $7 > 5$.

22. $-6 < 4$, $3 > 2$.

23. $-\frac{3}{4} > -\frac{14}{9}$, $\frac{3}{7} < \frac{8}{3}$.

24. $\frac{8}{5} > \frac{2}{3}$, $-\frac{7}{18} < -\frac{2}{9}$.

Solve the inequalities:

25a. $x + 4 > 2 - 3x$.

25b. $3 + 5x < 7x + 4$.

26a. $4(x - 1) > 2 + 7x$.

26b. $3(x - 2) < 4x - 9$.

27a. $\frac{3x}{2} - \frac{3}{5} < 4x - 3$.

27b. $\frac{x}{5} - 3\frac{1}{3} > 1\frac{3}{4} - \frac{5}{2}x$.

28a. $\frac{37-2x}{3} + 9 < \frac{3x-8}{4} - x$.

28b. $3 - \frac{3x}{2} > \frac{5}{8} - \frac{4x-3}{6}$.

29a. $(x - 1)^2 + 7 > (x + 4)^2$.

29b. $(1 + x)^2 + 3x^2 < (2x - 1)^2 + 7$.

30a. $\frac{7-6x}{2} + 12 < \frac{8x+1}{3} - 10x$.

30b. $8 + \frac{3x-4}{5} > \frac{x-1}{6} < \frac{5x-3}{8}$.

Determine for what values of x the expressions below are positive:

31. $2x - 16$.

32. $5 - 3x$.

33. $\frac{3}{8}x - 4$.

34. $\frac{x+1}{2} - 2x + 2\frac{1}{2}$.

35. $\frac{5-x}{8} + \frac{8-2x}{4}$.

Determine for what values of x the expressions below are negative:

36. $3x + 15$.

37. $7 - 14x$.

38. $5 - \frac{2}{3}x$.

39. $\frac{x-2}{3} + \frac{x}{2}$.

40. $\frac{3x-5}{2} - \frac{2x-1}{3} + 2$.

Solve the simultaneous inequalities:

41. $2x > 4x + 6$ and $4x + 3 < 2x + 1$.

42. $3x + 7 > 7x - 9$ and $x - 3 > -3x + 1$.

43. $5x - 3 > 1 + x$ and $\frac{1}{2} - 3x < \frac{2}{3}x - 5$.

44. $4x + 7 > 2x + 13$ and $3x - 18 < 2x + 1$.

45. $6x - 7 > 5x - 1$ and $3x + 6 > 8x - 4$.

46. $2(x - 3) - 1 < 5$ and $\frac{3x}{8} - 7 > \frac{x}{12}$.

47. $3x + 2 > x - 2$, $x + 15 > 6 - 2x$ and $x - 14 < 5x + 14$.

Determine for what values of a the fractions below are positive:

48a. $\frac{2a-3}{3a-2}$.

48b. $\frac{3a-8}{5-a}$.

49a. $\frac{2-3a}{2a+7}$.

49b. $\frac{3a-7}{2-5a}$.

Determine for what values of a the fractions below are negative:

50. $\frac{8-3a}{7a-2}$.

51. $\frac{5a+8}{3a-7}$.

52. On the basis of the inequality $(a-b)^2 > 0$, prove that the sum of the squares of two numbers is always greater than twice the product of the same numbers.

53. Prove that a proper fraction increases when the same positive number is added to its terms.

54. Prove that an improper fraction decreases when the same positive number is added to its terms.

55. Prove that the arithmetic mean of two numbers is greater than the geometric mean between them.

56. Prove that in any triangle the semiperimeter is greater than each of the sides.

57. Prove that in any non-isosceles right triangle the altitude dropped to the hypotenuse is less than half the hypotenuse.

58. Prove that in any right triangle the square of twice the altitude dropped to the hypotenuse is less than the sum of the square of the hypotenuse and twice the product of the legs.

59. Town N is located between two coal districts. In the first there is fatty bituminous coal with calorific value p (8000 calories), in the second lean coal with calorific value q (3500 calories); $q < p$. The cost of a ton of coal from district I, with loading, is a rubles (14), and delivery to N costs m rubles (16) per ton. For district II the costs are respectively b rubles (7) and n rubles (12). For what value of n is it more advantageous for town N to obtain coal from district II?

60. Let N rubles be the cost of an agricultural implement, P rubles the total cost of repair, n the number of possible uses of the implement without repair, and m the number of possible uses of the implement with repair. Under what condition will carrying out the repair pay for itself?

61. A vessel contains 4 L of water at room temperature, 13°. To what temperature must 5 L of water be heated so that, when mixed with the first, water is obtained with a temperature not less than 25° and not more than 30°?

62. Brass is an alloy of copper and zinc, whose specific gravities range respectively from 8.8 to 9.0 and from 6.85 to 7.2. A piece of brass weighs 400 g and in water experiences an upward pressure of 50 g. How much copper and zinc are in the alloy?

63. In determining the speed of sound, two observation points were chosen at a distance of 1000 m. The sound travelled with the wind in not less than $2\frac{1}{2}$ seconds, and against the wind in not more than 3 seconds. Find the upper bound of the speed of sound.

Chapter XXI

Solving Indeterminate Equations of the First Degree

Solve the following equations in integers by the method of substitution:

1. $x + 2y = 7$.

2. $y - 5x = 12$.

3. $3x - 5y = 0$.

4. $5x + 8y = 0$.

5. $2x + 3y = 13$.

6. $5y - 7x = 21$.

7. $7x + 13y = 71$.

8. $14x - 9y = 11$.

Solve the following equations in integers by the method of successive divisions:

9. $2x + 3y = 7$.

10. $3x - 4y = 11$.

11. $5x + 3y = 6$.

12. $7x - 4y = 3$.

13. $7x + 5y = 12$.

14. $5x - 11y = 4$.

15. $11x + 8y = 73$.

16. $11x - 7y = -31$.

Can the following equations be solved in positive integers:

17. $2x + 6y = 25$.

18. $6x + 11y = -48$.

19. $8x + 7y = 3$.

20. $9x - 6y = 17$.

21. $10x + 13y = 16$.

22. $13x - 15y = 45$.

23. $8x + 6y = 12$.

24. $15x - 10y = 25$.

Solve the following equations in positive integers:

25. $4x + 11y = 47$.

26. $12x - 7y = 45$.

27. $11x + 18y = 120$.

28. $15x - 49y = 11$.

29. $18x - 35y = 30$.

30. $45x + 27y = 117$.

31. $\frac{3x}{5} + \frac{2y}{3} = 37$.

32. $\frac{x+15y}{x-21} = -20$.

33. $\frac{3x-14}{2} = \frac{2y-0.5}{5}$.

Find the smallest positive numbers satisfying the following equations:

34. $17x - 29y = 100$.

35. $13x - 15y = 2$.

36. $52x + 64y = 388$.

37. $16x - 25y = 1$.

38. $41x - 36y = 187$.

39. $9x + 20y = 547$.

Solve the following systems of equations in positive integers:

40. $2x - 5y = 5,\ 2y - 3z = 1$.

41. $8x - 5y = 6,\ 7z + 3y = 13$.

42. $3x + y + z = 14;\ 5x + 3y + z = 28$.

43. $4x + y + 3z = 30,\ 7x + y + 6z = 51$.

44. $x = 5y + 3 = 11z + 7$.

45. $x + 2y + 3z = 20,\ 3x + 5y + 4z = 37$.

46. $2x + 14y - 7z = 341,\ 10x + 4y + 9z = 473$.

47. Decompose the number 200 into two summands, one of which is divisible without remainder by 7, and the other by 13.

48. In how many ways, and in what ways, can one pay 149 rubles for tickets costing 3 rubles and 5 rubles?

49. Find two numbers whose difference is 10, given that the minuend is a multiple of 8 and the subtrahend a multiple of 17.

50. In how many ways, and in what ways, can one weigh a load of 114 kg, having weights of 5 and 3 kg?

51. In how many ways, and in what ways, can one weigh a load of 87 kg, having weights of 5 and 2 kg?

52. To two artels of workers 330 rubles were issued. Each worker of the first artel received 16 rubles, and each worker of the second 9 rubles. How many workers were in each artel?

53. How many five-kopeck and two-kopeck (bronze) coins can be placed along a meter, assuming the diameter of the first equals 25 mm and the diameter of the second 18 mm?

54. The fraction $\frac{7}{18}$ equals the difference of two fractions, one of which has denominator 9 and the other 13. Find these fractions.

55. For what value of x does the fraction $\frac{5x-1}{12}$ become a positive even number?

56. Find the general form of numbers, multiples of 5, which when divided by 8 give a remainder of 1.

57. Find the general form of numbers, multiples of 7, which when divided by 5 give a remainder of 2.

58. For what value of x does the fraction $\frac{3-7x}{10}$ become a positive number divisible by 4 with remainder 3?

59. Find the general form of numbers which when divided by 3 give remainder 2, and when divided by 7 give remainder 3.

60. Find the general form of numbers which when divided by 7 give remainder 4, and when divided by 8 give remainder 3.

61a. A marksman receives 8 kopecks for each successful shot, and for each unsuccessful one himself pays 27 kopecks. Having made a certain number of shots, fewer than 120, he gained 97 kopecks. How many shots were successful and how many unsuccessful?

61b. A water pipe of length 105 m is being laid; in the warehouse there are pipes of 3 m and 4.5 m length. How many pipes, and of which kind, must be laid?

62. In the rotation of two meshing gear wheels, one of which has 19 teeth and the other 23, the first tooth of one wheel fell into the first gap of the other. How many complete revolutions must both wheels make for the first tooth to fall again into the first gap, how many for it to fall into the second gap, the third, and so on?

63. In the rotation of two meshing gear wheels, one of which has 25 teeth and the other 36, the first tooth of one wheel fell into the first gap of the other. How many complete revolutions must both wheels make for the first tooth to fall again into the first gap, how many for it to fall into the second gap, the third, and so on?

64. Find a three-digit number the sum of whose digits is 16; if 99 is added to this number, one obtains the number denoted by the same digits in reverse order.

65. Find the smallest of the numbers which when divided by 3, 4, 5 give remainders

1, 2, and 3.

66. Find the general form of numbers which, being multiples of 5, when divided by 8, 11, and 3 give remainders 1, 3, and 1.

67. Find the smallest of the numbers which when divided by 5, 6, 7, and 8 give remainders 3, 1, 0, and 5.

68. Pay 25 kopecks with coins of 2, 3, and 5 kopecks.

69. A factory shop of 75 workers was awarded a bonus of 200 rubles for exemplary work. It was decided to spend it on the purchase of theater tickets. How many tickets, and of which kind, will have to be bought, if the box office has tickets at 4 rubles, 2 rub. 75 kop., and 2 rubles?

Chapter XXII

Continued Fractions

Convert the following continued fractions into simple ones:

1. (2, 1, 2, 3, 2).

2. (2, 3, 1, 1, 12).

3. (0, 2, 1, 4, 3, 2).

4. (a, b, a, b, a).

5. $(0, x, 3x, x, 2x)$.

6. $(a-1, a, a+1, a)$.

Convert the following simple fractions into continued ones:

7a. $\frac{117}{55}$. **7b.** $\frac{151}{45}$.

8a. $\frac{117}{139}$. **8b.** $\frac{47}{64}$.

9a. $\frac{239}{99}$. **9b.** $\frac{137}{52}$.

10a. $\frac{71}{193}$. **10b.** $\frac{22}{7}$.

11. $\frac{a^4+2a^3-1}{a^3+a-1}$.

Find the convergents of the following continued fractions and compute the bounds of the errors of these approximations:

12a. $\frac{99}{239}$. **12b.** $\frac{55}{117}$.

13a. $\frac{55}{89}$. **13b.** $\frac{463}{640}$.

14a. $\frac{1702}{3919}$. **14b.** $\frac{685}{126}$.

Find the convergents of the infinite continued fractions and determine the bounds of their errors:

15a. $(1, 3, 5, 7, 9, 11 \ldots)$. **15b.** $(2, 4, 6, 8, 10, 12 \ldots)$.

16a. $(0, 10, 100, 1000 \ldots)$. **16b.** $(0, 5, 50 \ldots)$.

Convert the following roots into continued fractions:

17a. $\sqrt{2}$. **17b.** $\sqrt{5}$.

18a. $\sqrt{3}$. **18b.** $\sqrt{11}$.

19a. $\sqrt{20}$. **19b.** $\sqrt{12}$.

20a. $\sqrt{7}$. **20b.** $\sqrt{13}$.

21a. $\sqrt{a^2+1}$. **21b.** $\sqrt{a^2+2}$.

Convert the following continued fractions into irrational expressions:

22a. $(4, 8, 8, 8\ldots)$. **22b.** $(5, 10, 10, 10\ldots)$.

23a. $(3, 1, 6, 1, 6\ldots)$. **23b.** $(3, 2, 6, 2, 6\ldots)$.

24a. $(0, 2, 3, 2, 3, 2\ldots)$. **24b.** $(0, 1, 2, 1, 2\ldots)$.

25a. $(a, 2, 2a, 2, 2a\ldots)$. **25b.** $(a, 1, 2a, 1, 2a\ldots)$.

Solve the indeterminate equations in integers:

26a. $8x + 13y = 1$. **26b.** $7x + 12y = 1$.

27a. $9x - 14y = 3$. **27b.** $10x - 17y = 2$.

28a. $23x + 16y = 2$. **28b.** $41x + 29y = 1$.

29a. $7x - 11y = 1$. **29b.** $17x - 25y = 3$.

30a. $49x + 34y = 6$. **30b.** $29x + 17y = 25$.

Expand into continued fractions and compute approximately the following logarithms:

31a. $10^x = 500$. **31b.** $10^x = 800$.

32a. $10^x = 4000$. **32b.** $10^x = 60$.

33a. $72^x = 432$. **33b.** $36^x = 432$.

34a. $50^x = 500$. **34b.** $75^x = 375$.

Expand into continued fractions and compute approximately the roots of the equations:

35a. $x^3 - 2x - 5 = 0$. **35b.** $x^3 - x - 3 = 0$.

36a. $x^3 + x^2 + x - 1 = 0$. **36b.** $x^3 + x^2 + x - 2 = 0$.

37. At the end of the 19th century, the famous chemist D. I. Mendeleev, in the laboratory of the Chamber of Weights and Measures, found with great precision that 1 pound = 0.40951241 kg and 1 arshin = 0.711200 m. For rapid conversion, especially mental, these numerical ratios are awkward. Find three successive convergents expressing these ratios, and indicate the degree of accuracy.

38. The ratio of the circumference of a circle to its diameter is approximately 3.1415926535; expand it into a continued fraction and find five successive convergents for it, indicating the degree of accuracy of each.

Chapter XXIII

Investigation of Equations

Investigation of First-Degree Equations with One Unknown

Determine for what values of a the following equations have positive solutions:

1. $5(x-3)=3(3x-2a)$.

2. $3(x+1)=4+ax$.

3. $\frac{5}{3+x}=\frac{a}{x}$.

4. $\frac{3}{x+1}=8-a$.

Determine for what values of a the following equations have negative solutions:

5. $7-a=\frac{2}{x-1}$.

6. $\frac{3}{4x-a}=\frac{2}{ax-5}$.

Modify the following equations, which have negative solutions, so that their solutions become positive:

7. $4x-75=6(x-10)+85$.

8. $13x-22=17(x-2)+28$.

9. $5(3-7x)+4(3x-7)=35+x$.

10. $6(x-1)-12x=12(x+3)-2(x+5)$.

Investigate for what values of the literal quantities entering the following equations these equations have positive, negative, zero, infinite, and indeterminate solutions:

11. $\frac{a}{a-x}=\frac{m}{n}$.

12. $3ax+b=b(a+x)$.

13. $ax+m=b(x+n)$.

14. $\frac{px+m}{x+m}=\frac{a}{b}$.

15. Two parties of workers together received 120 rubles; each worker of the first party received 7 rubles, and each worker of the second 5 rubles; in the second party there were three workers more than in the first. How many workers were in each party?

16. Determine a two-digit number in which the number of tens is half the number of units, and the difference between the numbers of units and tens is 6.

Solve and investigate the following general problems leading to literal equations:

17. In one reservoir a buckets of water are poured, and in another b buckets. Each hour m buckets are added to the first and n buckets to the second. In how many hours will the numbers of buckets in both reservoirs become equal?

18. The father is a years old, the son b years. In how many years will the father be k times as old as the son?

19. What number must be subtracted from the numbers a and b so that the ratio of the differences equals k?

20. The front wheel of a cart has a circumference of a meters, the rear b meters. How long is the path on which the front wheel makes one revolution more than the rear?

21. What number must be added to the numerator and denominator of the fraction $\frac{a}{b}$ so that it becomes the fraction $\frac{m}{n}$?

22. In a liters of water b grams of salt are dissolved; how much water must be added so that each liter contains m grams of salt?

23. Divide the number a into two parts so that the sum of the products of the first part by m and of the second by n equals the sum of the products of the first part by p and of the second by q.

24. In triangle ABC the sides are given: $AB = c$, $AC = b$, $BC = a$. Drawing the bisector of the external angle at vertex C, mark the point D of its intersection with the extension of side AB. Determine the distance AD.

Investigation of a System of First-Degree Equations with Two Unknowns

25. Determine for what values of a the system of equations $x+y = a$ and $3x+2y = 10$ gives positive solutions.

26. Determine for what values of a the system of equations $4x - 3y = 6$ and $-5x + ay = 8$ gives negative solutions.

27. Determine the value of a for which the system of equations $3x - 7y = 15$ and $46x + ay = 60$ has no solutions.

28. Determine the value of a for which the system of equations $2x + 5y = 7$ and $7x - ay = 9$ has no solutions.

29. Determine the values of a and b for which the system of equations $ax - by = 15$

and $4x + by = 2$ has an infinite set of solutions.

30. Determine the values of a and b for which the system of equations $ax - y = b$ and $4x + 3y = 10$ has an infinite set of solutions.

Investigation of Second-Degree Equations

In the following problems determine the conditions under which the roots of the equations are real and positive, and investigate for the roots certain commensurable integral values corresponding to particular assumptions.

31. Find two numbers whose sum is a and whose product is b.

32. Inscribe in a given square with side a another square with side b.

33. Given the hypotenuse a, construct a right triangle equal in area to a square with side b.

34. A circle has radius R, and a point outside it is at distance d from its center. Draw through this point a secant whose interior segment is equal to the radius of the circle.

35. Inscribe in a circle of radius R a rectangle whose area is equal to the area of a square with side k.

In the following second-degree equations with two unknowns, determine those real values of x for which y is also real:

36. $x^2 + y^2 - 2xy + x = 0.$

37. $2x^2 - 2xy + y^2 + 2x - 4y + 1 = 0.$

Chapter XXIV

Limits

Find the limits:

1. $\lim_{n\to 0}\frac{5}{n+1}$.

2. $\lim_{x\to 0}\frac{x^2+6x-1}{5x+3}$.

3. $\lim_{x\to 2}\frac{x^2-3x+4}{x+4}$.

4. $\lim_{x\to 1}\frac{x^2-1}{x+1}$.

5. $\lim_{x\to 0}\frac{x^4+3x^2}{x^5+x^3+2x^2}$.

6. $\lim_{a\to 2}\frac{a^2-4}{a-2}$.

7. $\lim_{x\to 1}\frac{x^2-5x+4}{x^2+x-2}$.

8. $\lim_{x\to 3}\frac{x^2+2x-15}{x^2-9}$.

9. $\lim_{x\to a}\frac{x^3-a^3}{x^2-a^2}$.

10. $\lim_{x\to\infty}\frac{3x+1}{x+2}$.

11. $\lim_{x\to\infty}\frac{5x^2-1}{4x^2+3}$.

12. $\lim_{x\to\infty}\frac{2x^2+1}{x^3+3}$.

13. $\lim_{x\to\infty}\frac{ax+b}{cx+d}$.

14. $\lim_{x\to\infty}\frac{ax^3+b}{cx^3+d}$.

Find the limits of the sums of the terms of infinitely decreasing progressions:

15. $2+1+\frac{1}{2}+\frac{1}{4}+\ldots$

16. $1+\frac{1}{3}+\frac{1}{9}+\ldots$

17. $2-1+\frac{1}{2}-\frac{1}{4}+\ldots$

18. $2+\sqrt{2}+\ldots$

19. $0.3333\ldots$

20. $2.3(4)$.

If h approaches zero, which of the following quantities will be infinitesi-

mal?

21. $12h$, $\sqrt{h}$, $0.0001h$.

22. $\frac{1}{h}$, $\frac{0.01}{h}$, $\frac{h}{h^2}$, $\frac{h^2}{h}$, h^3.

23. 2^h, 10^h, $\lg h$.

24. $\sin h$, $\cos h$, $\operatorname{tg} h$, $\operatorname{ctg} h$.

To what limits do the following expressions tend as $x \to 0$?

25. $5x$, $ax+b$, $\frac{ax+b}{cx+d}$, $\frac{ax^2+b}{cx^2+d}$, $\frac{ax^2+b}{cx^3+d}$.

26. $a^x - 1$, $a^x - 2$, $\lg x$, $\lg(5-x)$, $\sin\left(\frac{ax+\pi}{bx+2}\right)$.

27. Find the limit of the product:

$$\sqrt{5}\,\sqrt[4]{5}\,\sqrt[8]{5}\,\sqrt[16]{5}\ldots\sqrt[2^n]{5}\ldots$$

Answers

The answers below are keyed to the numbered problems in this edition. Only entries for which an answer is supplied are listed. For a paired problem, an answer marked *a* belongs to variant a; no answer is supplied for variant b unless a separate b-entry is printed.

Chapter IX. Irrational Expressions

1. 4.
2. $-a^2$.
3. a^3.
4. a^3.
5. 6.
6. 6.
7. $\frac{a^2}{3}$.
8. $-\frac{a^2}{b^3}$.
9. a^4b^2c.
10. $-3a^4b$.
11. $\frac{1}{3}$.
12. $\frac{3}{2}$.
13. $\frac{1}{a^2}$.
14. $-\frac{1}{a^4}$.
15. -2.
16. $a^5\sqrt[n]{-1}$.
17. $\frac{2b^3}{a}$.
18. $\frac{5b^2}{2a^n}$.
19. $\frac{5a^3c^2m}{2}$.
20. $\frac{2a^{2n}b^4}{3}$.
21. $\frac{3a^{2n-1}b^6}{10c^2}$.
22. $-\frac{100}{a^{4n}b^{3m-1}}$.

23. $\frac{3a^2d}{2b^3c^4}$.

24. $\frac{28b^6d}{a^5c^3}$.

25. $\frac{2c^m}{a^4b^{n-3}f^{2n-1}}$.

26. $-\frac{3b^3q^2}{10a^mp^4f^a}$.

27. $\frac{16b^2(a+b)^3}{a^4}$.

28. $4a^3b^4c^2$.

29. $\frac{b^{p+5}}{9a^5c}$.

30. $\frac{4c^{n-2}}{a^{n-10}b^5ad^3}$.

31. $2\sqrt{2}$.

32. $5\sqrt{3}$.

33. $3\sqrt[3]{3}$.

34. $-3\sqrt[3]{4}$.

35. $2\sqrt[4]{3}$.

36. $5\sqrt[4]{2}$.

37. $3\sqrt[5]{2}$.

38. $-2\sqrt[5]{7}$.

39. $18\sqrt{5}$.

40. $2\sqrt[4]{3}$.

41. $a^2\sqrt[4]{c^3}$.

42. $a^3b\sqrt[5]{b}$.

43. $xy\sqrt[3]{xy^2}$.

44. $ab\sqrt[4]{ab^2}$.

45. $2a^2\sqrt{b}$.

46. $4x^2y\sqrt[3]{y}$.

47. $12c^2d\sqrt{5}$.

48. $a^2\sqrt{a}$.

49. $\frac{a^2\sqrt[3]{a^2}}{b^3}$.

50. $\frac{\sqrt[6]{a^5}}{b^3}$.

51. $\frac{3\sqrt{6z}}{10x}$.

52. $-\frac{9\sqrt[3]{m}}{10a^2}$.

53. $\frac{(a-b)\sqrt{y}}{5}$.

54. $\frac{\sqrt{a-b}}{b}$.

55. $\frac{(y^2-x^2)\sqrt[3]{y-x}}{2}$.

56. $\frac{b\sqrt[3]{a^2-b^2}}{a}$.

57. $2a^5bc^p\sqrt[m]{2b^nc}$.

58. $x^2y^4z^2\sqrt[2r]{xy^2z^5}$.

59. $\frac{3a^2b}{c^2}\sqrt[n]{\frac{9a^5c}{b}}$.

60. $\sqrt{12}$.

61. $\sqrt{180}$.

62. $\sqrt[3]{54}$.

63. $\sqrt[3]{375}$.

64. $\sqrt[5]{160}$.

65. $\sqrt{5a^2}$.

66. $\sqrt[4]{2x^4}$.

67. $\sqrt[4]{625a}$.

68. $\sqrt[3]{m^3n}$.

69. $\sqrt{an^4}$.

70. $\sqrt{9a^3x}$.

71. $\sqrt[3]{m^7n}$.

72. $\sqrt{\frac{a}{4}}$.

73. $\sqrt[3]{\frac{x^3}{y}}$.

74. $\sqrt[3]{-\frac{b}{a^2}}$.

75. $\sqrt[5]{m^5-1}$.

76. $\sqrt{\frac{m+n}{m-n}}$.

77. $\sqrt[3]{24a^4bc^{11}}$.

78. $\sqrt[m]{3^{m+1}a^{mn+2}b^{m+1}}$.

79. $\sqrt[4]{\frac{162a^{n+8}c^{12}}{b^3}}$.

80. $\sqrt[3]{a^2}$.

81. $ab\sqrt[4]{ab^2}$.

82. $b\sqrt[3]{a^2}$.

83. $\sqrt[n]{ab^2}$.

84. $b\sqrt[3]{3a^2}$.

85. $\sqrt[3]{3a^mb^2}$.

86. $\sqrt[6]{8a^2b^n}$.

87. $\sqrt[3n]{\frac{4a^5}{3b^8c^9}}$.

88. $\sqrt[4]{\frac{10c}{9b^3a^2}}$.

89. $\frac{a^2\sqrt{bc}}{b^3}$.

90. $\sqrt[12]{a^{10}}$ and $\sqrt[12]{a^9}$.

91. $\sqrt[6]{4a^4}$ and $\sqrt[6]{ab^5}$.

92. $\sqrt[12]{16a^8b^4}$ and $\sqrt[12]{27a^9b^3}$.

93. $\sqrt[36]{\frac{27a^{15}}{b^9}}$ and $\sqrt[36]{\frac{10000b^8}{a^4}}$.

94. $\sqrt[m^2n]{\frac{3^na^{2n}}{b^nc^{3a}}}$ and $\sqrt[m^2n]{\frac{2^ma^mb^{2n}}{c^{3n}}}$.

95. $\sqrt[24]{a^4b^6}$, $\sqrt[24]{a^6}$ and $\sqrt[24]{a^9}$.

96. $\sqrt[30]{a^{10}b^5}$, $\sqrt[30]{a^6b^8}$ and $\sqrt[30]{a^6b^{12}}$.

97. $\sqrt[30]{\frac{x^{15}}{y^{15}}}$, $\sqrt[30]{\frac{y^{18}}{z^{12}}}$ and $\sqrt[30]{\frac{a^{20}}{b^{10}}}$.

98. $\sqrt[4n]{\left(\frac{x-1}{x+1}\right)^n}$, $\sqrt[4n]{\left(\frac{x+1}{x-1}\right)^2}$ and $\sqrt[4n]{\left(\frac{x}{y}\right)^4}$.

99. $\sqrt[6n]{\left(\frac{a+b}{x}\right)^3}$, $\sqrt[6n]{\left(\frac{a}{x+y}\right)^n}$ and $\sqrt[6n]{\left(\frac{a}{b}\right)^2}$.

100. $3y\sqrt[3]{x^2y^2}$.

101. $\frac{a^2b\sqrt[3]{6abd}}{3cd}$.

102. $\frac{a}{c^2}\sqrt[3]{\frac{5ab}{2}}$.

103. $a\sqrt[3]{a^2-b^2}$.

104. $a\sqrt[4]{a-b}$.

105. $\frac{b\sqrt[3]{5a^2b^2n^2}}{n}$.

106. $\frac{b}{a^nc^{m-n+1}}\sqrt[m+n]{\frac{b^{5a}}{c^n}}$.

107. $\frac{a^3(a+b)\sqrt[3]{(a-b)^2}}{a-b}$.

108. $\frac{a(a-2b)\sqrt{ab}}{c^2}$.

109. $\sqrt{3}$ and $2\sqrt{3}$.

110. $3\sqrt{7}$ and $2\sqrt{7}$.

111. $\sqrt[3]{2}$ and $2\sqrt[3]{2}$.

112. $2\sqrt[4]{5}$ and $3\sqrt[4]{5}$.

113. $3\sqrt{2}$, $8\sqrt{2}$ and $4\sqrt{2}$.

114. $3\sqrt[3]{2}$, $2\sqrt[3]{2}$ and $6\sqrt[3]{2}$.

115. $\frac{2}{3}\sqrt{3}$ and $2\sqrt{3}$.

116. $\frac{\sqrt{10}}{5}$ and $\frac{\sqrt{10}}{15}$.

117. $\frac{\sqrt{5}}{20}$ and $\frac{\sqrt{5}}{5}$.

118. $\frac{2\sqrt[3]{9}}{3}$ and $\frac{\sqrt[3]{9}}{2}$.

119. $\frac{\sqrt{5}}{2}$ and $\frac{\sqrt{5}}{3}$.

120. $-\frac{\sqrt[3]{10}}{10}$ and $\frac{\sqrt[3]{10}}{12}$.

121. $a\sqrt[6]{ab}$ and $a^2b\sqrt[6]{ab}$.

122. $0.3\sqrt[3]{xy^2}$ and $\frac{0.4\sqrt[3]{x^2y^4}}{y}$.

123. $\frac{\sqrt{a^3-1}}{a}$ and $\frac{\sqrt{a^3-1}}{a^2}$.

124. $\frac{\sqrt{b(1-ab)}}{b}$ and $\frac{d\sqrt{b(1-ab)}}{c}$.

125. $(a-b)\sqrt{a-b}$, $\frac{(a^2+b^2)\sqrt{a-b}}{a-b}$ and $a\sqrt{a-b}$.

126. $\frac{x^2\sqrt{x-y}}{y}$, $\frac{x\sqrt{x-y}}{x-y}$ and $\frac{2\sqrt{x-y}}{y}$.

127. $2a\sqrt{a^2-2b^2}, b\sqrt[3]{a^2-2b^2}$ and $-\frac{\sqrt{a^2-2b^2}}{ab}$.

128. $\frac{y\sqrt[n]{x^3y}}{x}$, $\sqrt[n]{x^3y}$ and $\frac{(2x-y)\sqrt[n]{x^3y}}{x}$.

129. $2\left(4\sqrt{2}+\sqrt[3]{3}\right)$.

130. $4\left(2\sqrt[4]{7}-\sqrt[5]{3}\right)$.

131. $2(2b\sqrt{c}-a\sqrt{b})$.

132. $3c\sqrt[4]{d}-2a\sqrt[5]{b^4}$.

133. $-\sqrt{2}$.

134. $129\sqrt{5}$.

135. $22\sqrt[3]{5}$.

136. $-\frac{5\sqrt{5}+13\sqrt{2}}{3}$.

137. $\frac{28\sqrt{6}+9\sqrt[3]{2}-4\sqrt{11}}{4}$.

138. $\frac{63\sqrt{2}-8\sqrt{3}}{6}$.

139. $(a+b-3)\sqrt{a}$.

140. $(15a^2+3a-6)\sqrt[3]{a}$.

141. $7ab\sqrt{5a}$.

142. $-4a^2c\sqrt{3d}$.

143. $2y\sqrt[3]{x^2y^2}$.

144. $-2n\sqrt{m-n}$.

145. $-\frac{5}{2}\sqrt{2(2-x)}$.

146. 0.

147. $\frac{x(2x-1)\sqrt[4]{x-1}}{2}$.

148. $x^2\sqrt[3]{x^3-y^3}$.

149. 9.

150. $2\sqrt[3]{4}$.

151. $-\frac{15}{2}\sqrt[3]{4}$.

152. $\frac{1}{3}$.

153. -60.

154. $72\sqrt[4]{20}$.

155. $448 + \frac{16}{3}\sqrt{6}$.

156. 68.

157. $-33\sqrt{5}$.

158. $6\sqrt{2} - 3\sqrt{15} + 8\sqrt{3} - 6\sqrt{10}$.

159. 84.

160. $-\sqrt{2}$.

161. $a^4b\sqrt{b}$.

162. $2a\sqrt[3]{x^2}$.

163. $30a^3\sqrt[3]{3}$.

164. $\frac{12b}{a}$.

165. $\frac{a\sqrt[3]{x}}{2c}$.

166. $4a^2x\sqrt[4]{2}$.

167. $-ab^5\sqrt[3]{25}$.

168. $\frac{a(b+\sqrt{b}-1)}{b}$.

169. $a\sqrt{b} + b - \frac{a\sqrt{b}}{b} - 1$.

170. $a\sqrt[3]{b} - b\sqrt[3]{a}$.

171. $\sqrt[6]{108}$.

172. $\frac{1}{\sqrt[15]{144}}$.

173. 6.

174. $\sqrt[36]{1152}$.

175. $3\sqrt[4]{200} - 2\sqrt[12]{2048} + 6\sqrt[12]{5000}$.

176. $2\sqrt[28]{11^{11}} + 3\sqrt[4]{40} - 4\sqrt[12]{625000}$.

177. $2\left(3 - \sqrt[6]{72} - 4\sqrt[3]{9}\right)$.

178. $11\sqrt[3]{4} - 15\sqrt[6]{2}$.

179. $\sqrt[12]{a^{11}b^{11}}$.

180. $15a^3b^2\sqrt[6]{108a^4b^3c^5}$.

181. $a^8b^6\sqrt[4]{a^3b}$.

182. $a^2\left(a\sqrt[6]{ab^3} - 2b\sqrt{a} - ab\sqrt[6]{a^3b^2}\right)$.

183. $2a^2\left(a - \sqrt[15]{a} - a^2\sqrt[6]{a}\right)$.

184. $\sqrt[6]{a^5} - a + a\sqrt[15]{a^2} - \sqrt[30]{a^{23}} + \sqrt[15]{a^{14}} - a\sqrt[15]{a}$.

185. 2.

186. 3.

187. $\frac{2\sqrt{3}}{7}$.

188. $2\sqrt[3]{2}$.

189. $\frac{2}{3}\left(5 - 3\sqrt[3]{20} + 15\sqrt[3]{4}\right)$.

190. $\sqrt{5}$.

191. $a^2\sqrt[3]{2}$.

192. $3\sqrt[4]{a}$.

193. $a\sqrt{\frac{2b}{3}}$.

194. $ab\sqrt{b} - \sqrt{x}$.

195. $\sqrt{a} - \sqrt[4]{a^2x} - \frac{4a}{x}\sqrt[4]{x^3}$.

196. $2x\sqrt[4]{y^3} - 3y\sqrt[4]{\frac{x^2y}{2}} + \sqrt{y}$.

197. $\sqrt[3]{a} - \sqrt[3]{b}$.

198. $\sqrt[3]{ab} - \sqrt[3]{2b^2}$.

199. $\sqrt{2a} + \sqrt[4]{6ab^2} + b\sqrt{3}$.

200. $a\sqrt{a} - \sqrt[4]{2a^3b^3} + b\sqrt{2b}$.

201. $x\sqrt[3]{x} - \sqrt[3]{x^2y^2} + y\sqrt[3]{y}$.

202. $\sqrt[6]{3}$.

203. $2\sqrt[5]{2500}$.

204. $2\left(\frac{1}{\sqrt[4]{6}} - \sqrt{2} + \frac{1}{\sqrt[6]{6}}\right)$.

205. $\frac{1}{\sqrt[6]{a}}$.

206. $\sqrt[6]{8a}$.

207. $a^4\sqrt{2}$.

208. $10\sqrt[6]{a^5}$.

209. $\frac{3}{2a^2}\sqrt[6]{432ab^5}$.

210. $x\sqrt[3]{\frac{5x^2}{y}}$.

211. $\frac{6a^2b^2}{d}\sqrt[15]{\frac{abd^{10}}{c}}$.

212. $\frac{ab+x^2}{x}\sqrt[6n]{\frac{a^{n+5}}{x^{10}}}$.

213. $3\sqrt{\frac{x+y}{x-y}}$.

214. $\frac{x(x^2-y^2)}{2a^2}\sqrt[3]{4a^2(x+y)^2}$.

215. $ab^2\sqrt{2a} - ab\sqrt[12]{8a^8b^7} + a^2b^2$.

216. $\sqrt[5]{4x^2} + \sqrt[5]{2x} \cdot \sqrt{3} + 3$.

217. $2\sqrt[3]{a^2x} + \sqrt{ax}$.

218. $x\sqrt{3xy} - x\sqrt[4]{12xy^3} + 2xy$.

219. a^3.

220. $a\sqrt[3]{a}$.

221. $2x^3\sqrt[4]{2x^3}$.

222. $-a^8b^2\sqrt[8]{a^6b^5}$.

223. $3a^{10}x^5\sqrt[3]{3a^2x}$.

224. $16a\sqrt[3]{9a}$.

225. $(x-y)\sqrt[5]{(x-y)^3}$.

226. $\frac{b^7\sqrt[4]{ab^2}}{a^4}$.

227. $\frac{a^2b^3}{\sqrt[3]{16a^{2n}b}}$.

228. $(x^2+y^2)^{mp}$.

229. $5-2\sqrt{6}$.

230. $\frac{33}{4}+2\sqrt{2}$.

231. $2\sqrt[3]{2}+4\sqrt[6]{2}+2$.

232. $3\sqrt{3}-18\sqrt[3]{2}+12\sqrt[6]{432}-16$.

233. $11-2\sqrt{6}+4\sqrt{3}-6\sqrt{2}$.

234. $48-12\sqrt{10}-12\sqrt{5}+20\sqrt{2}$.

235. 10.

236. 8.

237. 16.

238. $\frac{a^2b^3-16ab\sqrt{b}+64}{16a}$.

239. $a^4\sqrt{a}(7+5\sqrt{2})$.

240. $\sqrt[3]{a}$.

241. $\sqrt[15]{a^4}$.

242. $\sqrt{5}$.

243. $2a\sqrt[4]{a}$.

244. $\sqrt[8]{a^7}$.

245. $\sqrt[6]{a^5}$.

246. $ac\sqrt[4]{ab}$.

247. $\sqrt[12]{a^4b^3}$.

248. $\sqrt[24]{x^{41}}$.

249. $\sqrt[12]{\frac{x^9}{y}}$.

250. $\sqrt[24]{2^8\cdot 3^3x^{11}y^7}$.

251. 12.

252. 3.

253. 2.

254. 4.

255. $\sqrt{a}$.

256. $\frac{\sqrt{m}}{m}$.

257. $\sqrt[3]{a}$.

258. $\frac{(m+n)\sqrt{m-n}}{m-n}$.

259. $2\sqrt{2}$.

260. $3\sqrt[4]{2}$.

261. $\frac{\sqrt[3]{9}}{3}$.

262. $(a+b)\sqrt[3]{(a-b)^2}$.

263. $\frac{\sqrt[3]{(a^2-b^2)^2}}{a+b}$.

264. $\frac{a-2\sqrt{ab}+b}{a-b}$.

265. $\frac{a(1+\sqrt{a})}{1-a}$.

266. $2\sqrt{3}-3$.

267. $\sqrt{(1-a)(1+\sqrt{a})}$.

268. $\frac{n\left(\sqrt[3]{a^2}+\sqrt[3]{ab}+\sqrt[3]{b^2}\right)}{a-b}$.

269. $3(5\sqrt{2}-6+3\sqrt{6}-4\sqrt{3})$.

270. $\frac{\sqrt{5}+\sqrt{6}+\sqrt{7}}{2}$.

271. $\sqrt{3}-\sqrt{2}$.

272. $n\sqrt{\sqrt{2}+\sqrt[3]{3}}\left(\sqrt[3]{3}-\sqrt{2}\right)\left(4+2\sqrt[3]{9}+3\sqrt[3]{3}\right)$.

273. $-\frac{a^2+\sqrt{a^4-b^4}}{b^2}$.

274. $\left(\sqrt{2}-\sqrt[4]{3}\right)(2+\sqrt{3})$.

275. $\frac{\left(2\sqrt{3}+\sqrt[4]{3}\right)\left(12+\sqrt{3}\right)}{3}$.

276. $\frac{\sqrt{6}+\sqrt{2}}{2}$.

277. $2+\sqrt{2}$.

278. $\frac{\sqrt{14}-\sqrt{6}}{2}$.

279. $2+\sqrt{3}$.

280. $\sqrt[4]{18}+\sqrt[4]{2}$.

281. $\sqrt[4]{45}-\sqrt[4]{5}$.

282. $\frac{\sqrt{10}+\sqrt{2}}{2}$.

283. $3+\sqrt{2}$.

284. $\sqrt{a}-\sqrt{b}$.

285. $\sqrt{a^2+b}+\sqrt{a^2-b}$.

286. $\frac{\sqrt{3}}{2}+\frac{\sqrt{5}}{5}$.

287. $\sqrt{0.315}+\sqrt{0.065}$.

288. $2\sqrt{2a}$.

289. $\frac{2\sqrt{5}}{5}$.

290. 1.

291. 14.

292. 2.

293. $12+5\sqrt{3}$.

294. $\frac{12-\sqrt{5}+\sqrt{29}}{20}$.

295. $4+\sqrt{11}-3\sqrt{7}$.

296. 6.

297. $\frac{6\sqrt{6(x-6)}}{x-6}$.

298. $\sqrt{a^2-b^2}$.

300. $\frac{4a\sqrt{a^2-x^2}}{x^2}$.

301. $\sqrt[4]{63}$.

302. $-2x\sqrt[8]{x^7}$.

303. a.

304. $\frac{a\sqrt[4]{a}}{2\sqrt[4]{x}}$.

305. $\frac{a+\sqrt{a^2-x^2}}{x}$.

306. $2a\sqrt[8]{a^7}$.

307. $31a\cdot\sqrt{ax}$.

308. $\frac{a^2}{2x^2}\sqrt[4]{ax^3}$.

309. $\frac{2}{\sqrt[3]{\frac{9a^4}{1-a}}}$.

310. $\frac{1}{\sqrt{1+x}}$.

311. $\frac{\sqrt{ax-n^2}}{a^2\sqrt[4]{ax}-\sqrt{an}}$.

312. $\frac{a}{c}-\frac{c}{b}$.

313. $\sqrt{2}$.

314. 1.

315. -1.

316. $a+b$.

317. $\frac{a-\sqrt{2ab-b^2}}{a}$.

318. $a^{\frac{2}{3}}$.

319. $a^{-\frac{3}{4}}$.

320. $a^{-\frac{3}{5}}b^{\frac{4}{5}}$.

321. $a^{-\frac{3}{2}}$.

322. $(a^2+b^2)^{\frac{1}{2}}$.

323. $\frac{b^{\frac{2}{3}}}{a^{\frac{1}{6}}(a^3-b^3)^{\frac{1}{3}}}$.

324. $x^{\frac{n+1}{m}}$.

325. $a^{\frac{1}{m}}$.

326. $x^{\frac{2}{m}}y^{-\frac{n}{m}}$.

327. $a^{-\frac{n}{r}}$.

328. $a^{\frac{1}{3}}b^{\frac{1}{15}}$.

329. $(a-x)^{\frac{2}{3}}$.

330. $\sqrt[6]{a^5}$.

331. $\sqrt[4]{a^{-3}}$.

332. $\sqrt[3]{(a+b)^2}$.

333. $3\sqrt{a}\sqrt[8]{(a-b)^3}$.

334. $\frac{1}{\sqrt[4]{4a}}$.

335. $\sqrt[4]{a+b}$.

336. $\frac{1}{\sqrt{mx}}$.

337. $\frac{1}{\sqrt[12]{x^{2n-3}}}$.

338. $\frac{1}{\sqrt[x]{e}}$.

339. $n^2\sqrt{n}$.

340. ± 2.

341. 27.

342. $\frac{1}{32}$.

343. 4.

344. $\frac{6}{5}$.

345. $\frac{4}{9}$.

346. 0.8.

347. $\frac{1}{27}$.

348. 5.

349. -52.

350. $ab\sqrt[60]{a^{25}b^{16}}$.

351. $\sqrt[12]{\frac{b}{a}}$.

352. $a+\sqrt{ab}+b$.

353. $a+b+\sqrt{ab}$.

354. $a^n-\sqrt{\frac{a^n}{b^n}}+\frac{1}{b^n}$.

355. $\sqrt[3]{a^2} - 2\sqrt[3]{ab} + 4\sqrt[3]{b^2}$.

356. $\sqrt[4]{a} - \sqrt[4]{b} + \sqrt[4]{c}$.

366. i.

367. $-i$.

368. 1.

369. 1.

370. i.

371. -1.

372. -1.

373. $-i$.

374. i.

375. $2i$.

376. $9i$.

377. ai.

378. $b^3 i$.

379. $\frac{3i}{2}$.

380. $\frac{a^2 i}{b^4}$.

381. $i\sqrt{a}$.

382. $3i\sqrt{x}$.

383. $i\sqrt{a^2 + c^2}$.

384. $(x - y)i$.

385. $5i$.

386. $2(5\sqrt{3} - 3)i$.

387. $4 + 17i$.

388. $5a - 2bi$.

389. -12.

390. $-\sqrt{ab}$.

391. $-x$.

392. $(a - b)i$.

393. $1 - 46i$.

394. $100 - 13i\sqrt{7}$.

395. $a + 3b + 2i\sqrt{ab}$.

396. -17.

397. $-i\sqrt{a}$.

398. $\sqrt{a}$.

399. $a + bi$.

400. $\frac{x(x-y)}{x^2+y^2} - \frac{y(x-y)}{x^2+y^2}i$.

401. $1 - i\sqrt{3}$.

402. $-\frac{191}{209} - \frac{60\sqrt{2}}{209}i$.

403. $3 - 5i\sqrt{2}$.

404. $-\frac{7\sqrt{3}-6}{30} - \frac{(3+2\sqrt{3})\sqrt{7}}{30}i$.

405. $a^2 - b^2 + 2abi$.

406. $7 - 6i\sqrt{2}$.

407. $\frac{-1+i\sqrt{3}}{2}$.

408. $-12\sqrt{5} - 49$.

409. $-2(7 + 6\sqrt{-2})$.

410. $-\frac{7+4\sqrt{-2}}{4}$.

411. $a^3 - 3ab^2 - (3a^2b - b^3)i$.

412. $+9 + 25\sqrt{-2}$.

413. $-i(\sqrt{3} - 2)^3$.

414. 1.

415. $2 + i$.

416. $1 - 2i$.

417. $2 + i\sqrt{3}$.

418. $\frac{3\sqrt{2}}{2} - i\frac{\sqrt{10}}{2}$.

419. $\sqrt{22} - i\sqrt{2}$.

420. $\frac{1}{2}(\sqrt{26} + i\sqrt{2})$.

421. $\frac{1+i}{\sqrt{2}}$.

422. $\frac{\sqrt[4]{2}}{2}(\sqrt{\sqrt{2}+1} + i\sqrt{\sqrt{2}-1})$.

Chapter X. Functions and Their Graphs

4. On the bisector 1st and 3rd quadrants.

5. On the bisector 2nd and 4th quadrants.

30. At a distance 32 km after 2 h 40 m.

31. Approximately starting from $d = 7.15$.

Chapter XI. Quadratic Equations

1a. 0 and $-\frac{b}{a}$.

2. 0 and $a + b$.

3. 0 and $\frac{1}{m+n}$.

4. 0 and $\frac{3a}{2}$.

5. $\pm 3ab$.

6. 0 and $2a$.

7. $\pm\sqrt{ab}$.

8. $\pm n$.

9. $\pm\frac{a+1}{a}$.

10. $\pm\sqrt{(a^2-ab+b^2)}$.

11. $\pm\sqrt{4a^2+b^2}$.

12. $-a$.

13. $3a$ and a.

14. $-7a^3$ and $5a^3$.

15. $a\pm b$.

16. $a-5b$ and $3b-a$.

17. $2a$ and $-\frac{a}{2}$.

18. $-\frac{a}{3}$ and $-\frac{a}{2}$.

19. $-\frac{3a}{b}$ and $-\frac{a}{3b}$.

20. $\frac{5a}{4b}$ and $-\frac{4a}{5b}$.

21. $\frac{m}{n}$ and $-\frac{n}{m}$.

22. $\frac{a}{b}$ and $\frac{b}{a}$.

23. $\frac{a}{b}$ and -1.

24. $\frac{a}{a+b}$ and $\frac{b}{a-b}$.

25. $\frac{a}{b}$ and $-\frac{b}{a}$.

26. $\frac{2a}{3}$ and $-\frac{5a}{7}$.

27. $\frac{3a-b\pm\sqrt{9a^2-10ab+b^2}}{2}$.

28. $a\pm 2b$.

29. $-\frac{a}{2}(3\pm\sqrt{3})$.

30. $\frac{a}{2}$ and $-\frac{5a}{6}$.

31. $-a$ and $-b$.

32. 1 and 1.

33. $\frac{ab}{a\pm b}$.

34. $\frac{2c}{a+b}$ and $-\frac{c}{a+b}$.

35. a and b.

36. a and b.

37. $\frac{a+b}{2(a-b)^2}(a^2+b^2\pm\sqrt{a^4-4a^3b+10a^2b^2-4ab^3+b^4})$.

38. $\frac{1}{3}(a+b+c\pm\sqrt{a^2+b^2+c^2-ab-ac-bc})$.

39. $\frac{a\pm\sqrt{a^2-4b(a-c)}}{2}$.

40. $\frac{5a+3b}{8}$ and $\frac{3a+5b}{8}$.

41. $-a$ and $\frac{a(c+1)}{c(2c+3)}$.

42. $\frac{ab+ac+bc\pm\sqrt{a^2b^2+a^2c^2+b^2c^2-a^2bc-ab^2c-abc^2}}{a+b+c}$.

43. a^n and a^{-n}.

44. a^{3n} and $-a^n$.

45. $\sqrt{a}$ and $-\sqrt{2a}$.

66. $(x-3)(x-4)$.

67. $(x-9)(x+12)$.

68. $(3x-2)(2x+3)$.

69. $(5x+2)(6x+5)$.

70. $[x-(3+i\sqrt{2})][x-(3-i\sqrt{2})]$.

71. $(x+11)(x+4)$.

72. $(x-3a)(x+2a)$.

74. $(x-a-\sqrt{b})(x+\sqrt{b})$.

75. $(x\sqrt{ab}-a-b)(x\sqrt{ab}-a+b)$.

76. $x^2-5x+6=0$.

77. $x^2-2x-24=0$.

78. $x^2+5x=0$.

79. $x^2-9=0$.

80. $8x^2-2x-1=0$.

81. $6x^2+13x+6=0$.

82. $x^2-(\sqrt{6}-\sqrt{3})x-3\sqrt{2}=0$.

83. $x^2-8x+13=0$.

84. $x^2+6x+24=0$.

85. $x^2-2x+11=0$.

86. $x^2+(2b-3a)x-6ab=0$.

87. $x^2+3(b-a)x+2a^2-5ab+2b^2=0$.

88. $6x^2-ax-a^2=0$.

89. $x^2-2ax+a^2-b^2=0$.

90. $x^2-\left(\frac{a}{b}+\frac{b}{a}\right)x+1=0$.

91. $(a+b)x^2-2ax+a-b=0$.

92. $(1-a)(1-b)x^2-[a(1-a)+b(1-b)]x+ab=0$.

93. $x^2-2ax+a^2-b=0$.

94. $x^2-2\sqrt{a}x+a+b=0$.

95. $qx^2+px+1=0$.

96. $ax^2+mbx+m^2c=0$.

97. $4x^2-p^2+4q=0$.

98. $a^2x^2+a(b-c)x-bc=0$.

99. $\frac{b(3ac-b^2)}{a^3}$.

100. $p^2 - q$.

101. ± 20.

103. $c \leqslant 27$ and $c > 27$.

106. -16.

107. 4.

108. $mp^2 - (m+1)^2 q = 0$.

109. $\frac{2q}{-p \pm \sqrt{p^2-4q}}$.

110. $\pm\sqrt{pq}$ and $\pm\sqrt{\frac{p}{q}}$.

111. $\frac{b+\sqrt{b^2+4ac}}{2}$ and $\frac{-b+\sqrt{b^2+4ac}}{2a}$.

112. $\frac{\pm b \pm \sqrt{b^2+4a}}{2}$.

113. $p\sqrt{\frac{S}{p^2+q^2}}$ and $q\sqrt{\frac{S}{p^2+q^2}}$.

114. $\frac{a-\sqrt{a^2-4}}{2}$, $(a \geqslant 2)$.

115. $\frac{a-100(b+c)+a\sqrt{[100(b+c)-a]^2+400ab}}{2}$.

116. $\frac{\sqrt{5}-1}{2}a$ and $\frac{3-\sqrt{5}}{2}a$.

117. The segment should be bisected.

118. The sides of the rectangle are: $\frac{p \pm \sqrt{p^2-4S}}{2}$; it is a square if $p^2 = 4S$.

119. $\frac{2\sqrt{5n^2-m^2}+4m}{5}$ and $\frac{2\sqrt{5n^2-m^2}-m}{5}$.

120. $\frac{c \pm \sqrt{c^2-4h^2}}{2}$ $(h \leqslant \frac{c}{2})$.

121. $\frac{a}{\sqrt{n}-1}$.

122. $\frac{d}{\sqrt{V_1^2+V_2^2}}$.

123. $\frac{mP}{\sqrt{m^2-mn+n^2}}$ and $\frac{nP}{\sqrt{m^2-mn+n^2}}$.

124. $\frac{\sqrt{n}}{\sqrt{m}-\sqrt{n}}h$.

125. $\frac{d\sqrt{n}}{\sqrt{n}+1}$.

126. $\sqrt{a^2+b^2}$.

127. Equation: $\sqrt{\frac{2x}{98}} + \frac{x}{330} = n$.

128. $\frac{a+b \pm \sqrt{a^2-ab+b^2}}{6}$; $\frac{a+b \pm \sqrt{(a-b)^2+4S}}{4}$.

129. $\frac{\sqrt{bc(bc+4a)}-bc}{2c}$ km per hour.

130. $\frac{\sqrt{mn(mn+4p)} \pm mn}{2n}$.

131. $\frac{2T \pm t + \sqrt{4T^2+t^2}}{2}$.

147. min. V at $t = 4$.

148. max. S at $t = 12.7$.

Chapter XII. Equations of Higher Degrees

1. $\pm 2;\ \pm 1.$

2. $\pm 2i;\ \pm 2i\sqrt{2}.$

3. $\pm\frac{2}{\sqrt{5}};\ \pm i.$

4. $\pm\sqrt{\frac{2}{3}};\ \pm\frac{i\sqrt{3}}{2}.$

5. $\pm\sqrt{a};\ \pm i\sqrt{b}.$

6. $x_{1.2} = \pm 2;\ x_{3.4} = \pm ai.$

7. $x_{1.2} = \pm\frac{a}{b};\ x_{3.4} = \pm\frac{b}{a}.$

8. Sum $= 0.$

9. $4\cdot(x-2)(x+2)\left(x-\frac{1}{2}\right)\left(x+\frac{1}{2}\right).$

10. $x^4 - 10x^2 + 9 = 0.$

11. $-1;\ \frac{1\pm i\sqrt{3}}{2}.$

12. $2;\ -1\pm i\sqrt{3}.$

13. $a;\ a\cdot\frac{-1\pm i\sqrt{3}}{2}.$

14. $\pm 2;\ \pm 2i.$

15. $\pm\frac{3\sqrt{2}}{2}(1\pm i).$

16. $\pm 2;\ \pm 1\pm i\sqrt{3}.$

17. $-\frac{2}{5};\ \frac{1\pm i\sqrt{3}}{5}.$

18. $\frac{\pm 1\pm i}{3}.$

19. $\pm\frac{b}{a};\ \pm\frac{b}{a}i;\ \frac{b\sqrt{2}}{2a}(\pm 1\pm i).$

20. $1;\ \frac{-1\pm i\sqrt{3}}{2};\ \sqrt[3]{2};\ \frac{-1\pm i\sqrt{3}}{\sqrt[3]{4}}.$

21. $1;\ \frac{-1\pm i\sqrt{3}}{2};\ 3;\ \frac{3(-1\pm i\sqrt{3})}{2}.$

22. $-1;\ \frac{1\pm i\sqrt{3}}{2};\ -\frac{1}{\sqrt[3]{3}};\ \frac{1\pm i\sqrt{3}}{2\sqrt[3]{3}}.$

23. $\frac{1}{2};\ \frac{-1\pm i\sqrt{3}}{4};\ -\frac{1}{\sqrt[3]{5}};\ \frac{1\pm i\sqrt{3}}{2\sqrt[3]{5}}.$

24. $32;\ 16(-1\pm i\sqrt{3});\ 1;\ \frac{-1\pm i\sqrt{3}}{2}.$

25. $-1;\ \frac{1\pm i\sqrt{3}}{2}\sqrt[3]{7^5};\ \sqrt[3]{7^5};\ \frac{-1\pm i\sqrt{3}}{2}.$

26. $\frac{3}{a};\ \frac{3(-1\pm i\sqrt{3})}{2a};\ -\frac{1}{a};\ \frac{1\pm i\sqrt{3}}{2a}.$

27. $1;\ \frac{\sqrt{5}-1\pm i\sqrt{10+2\sqrt{5}}}{4};\ \frac{-(\sqrt{5}+1)\pm i\sqrt{10-2\sqrt{5}}}{4};\ \sqrt[5]{9};\ \frac{\sqrt[5]{9}}{4}(\sqrt{5}-1\pm i\sqrt{10+2\sqrt{5}});$
$\frac{\sqrt[5]{9}}{4}(-\sqrt{5}-1\pm i\sqrt{10-2\sqrt{5}}).$

28. $0;\ \frac{7}{2}.$

29. $\frac{8}{5};\ -\frac{7}{4};\ -9.$

30. $2;\ -1;\ -1.$

31. $1;\ 2;\ -3.$

32. $-1;\ \pm 1.$

33. $5;\ \pm 1.$

34. $-3;\ \frac{1\pm i\sqrt{3}}{2}.$

35. $-6;\ -1 \pm i\sqrt{2}.$

36. $1;\ -2;\ \pm i\sqrt{2}.$

37. $1;\ -2;\ 3;\ -4.$

38. $2;\ -3;\ \frac{3\pm\sqrt{5}}{2}.$

39. $-1;\ \frac{n\pm i\sqrt{n(3n+4)}}{2n}.$

40. $a;\ a(1 \pm 2i\sqrt{3}).$

41. $-a;\ \frac{a\pm\sqrt{a^2-4b^2}}{2}.$

43. $1;\ \pm i.$

44. $\pm 1;\ 2;\ \frac{1}{2}.$

45. $\frac{3}{2};\ -\frac{2}{3};\ \pm i.$

46. $\pm 1;\ \frac{b\pm\sqrt{b^2-4a^2}}{2a}.$

47. $x_1 = 2;\ x_2 = \frac{1}{2};\ x_3 = -\frac{1}{3};\ x_4 = -3.$

48. $x_1 = 2;\ x_2 = \frac{1}{2};\ x_3 = \frac{-3+\sqrt{5}}{2};\ x_4 = \frac{-3-\sqrt{5}}{2}.$

49. $x_{1.2} = 4 \pm \sqrt{17};\ x_{3.4} = \frac{1\pm\sqrt{65}}{8}.$

50. $x_1 = 2;\ x_2 = -\frac{1}{2};\ x_3 = \frac{1}{3};\ x_4 = -3.$

51. $x_1 = -1;\ x_{2.3} = -2 \pm \sqrt{3};\ x_4 = 2;\ x_5 = \frac{1}{2}.$

54. $x_1 = 1;\ x_2 = x_3 = x_4 = -1;\ x_5 = 2;\ x_6 = \frac{1}{2}.$

56. $2(x-1)^2(x-2)\left(x-\frac{1}{2}\right).$

57. $6x^4 - 35x^3 + 62x^2 - 35x + 6 = 0.$

Chapter XIII. Irrational Equations

1. 2.

2. 20.

3. $-1.$

4. 7.

5. 6.

6. 7.

7. 4.

8. 4.

9. 0; 2.

10. $0;\ \frac{5}{2}.$

11. 2.
12. 2.
13. ± 2.
14. $3; -\frac{2}{3}$.
15. 81.
16. 5.
17. $2; \frac{5}{2}$.
18. $4; -\frac{10}{27}$.
19. $0; \frac{25}{16}$.
20. $-\frac{2}{3}$.
21. $\frac{2}{3}$.
22. 22.
23. $1; -\frac{27}{8}$.
24. 84; 19.
25. $\pm 3\sqrt{2}$.
26. $0; -5$.
27. $\frac{a}{4}$.
28. $0; a$.
29. $\pm\frac{1}{2}\sqrt{4a^2+2b^2}$.
30. $\frac{a}{b}; \frac{c}{d}$.
31. $-\frac{2a}{3}$.
32. $\frac{3a^2}{4}$.
33. $2a$.
34. $0; \pm a$.
35. $\frac{1\pm\sqrt{1+4b^2}}{2a}$.
36. $\frac{a\pm\sqrt{a^2+24ab}}{6}$.
37. $\frac{a+b}{2} \pm \frac{a-b}{4}\sqrt{2}$.

Chapter XIV. Systems of Equations of Degree Higher than the First

1. $6; 2$ and $-\frac{22}{3}; -\frac{14}{3}$.
2. $3; 5$ and $-3; -4$.
3. $1; 3$ and $19; -3$.
4. $4; 2$ and $3; 1$.
5. $1; \frac{1}{2}$ and $-1; -\frac{1}{2}$.
6. $11; 9$ and $-11; -9$.
7. $15; 3$ and $9.6; -2.4$.

8. 20; 100 and 100; 20.

9. 5.6; 5.2 and −7.6; 0.8.

10. 3; 4 and 1; 2.

11. $3a$; $2a$ and $-\frac{27}{16}a$; $-\frac{43}{16}a$.

12. $\frac{a}{a-1}$; $\frac{a}{a+1}$ and $-\frac{a}{1+a}$; $\frac{a}{1-a}$.

13. a; 0 and 0; b.

14. $\frac{a}{2}\left(1+\sqrt{\frac{b-2}{b+2}}\right)$; $\frac{a}{2}\left(1-\sqrt{\frac{b-2}{b+2}}\right)$ and $\frac{a}{2}\left(1-\sqrt{\frac{b-2}{b+2}}\right)$; $\frac{a}{2}\left(1+\sqrt{\frac{b-2}{b+2}}\right)$.

15. $\sqrt{ab}$; $\sqrt{\frac{a}{b}}$ and $-\sqrt{ab}$; $-\sqrt{\frac{a}{b}}$.

16. (3; 1), (−3; −1), $\left(12; -\frac{7}{2}\right)$ and $\left(-12; \frac{7}{2}\right)$.

17. (0; 0), (2; 2), $(\sqrt{2}; 2-\sqrt{2})$ and $(-\sqrt{2}; 2+\sqrt{2})$.

18. 3; 5 and 2; 1.

19. (3; 2), (−3; −2), $\left(\frac{8\sqrt{3}}{3}; -\frac{7\sqrt{3}}{3}\right)$ and $\left(-\frac{8\sqrt{3}}{3}; \frac{7\sqrt{3}}{3}\right)$.

20. (3; 5), (−3; −5) and (−8; 5).

21. (6; 5), (−6; 5), (6; −5) and (−6; −5).

22. $(4;\ 2\sqrt{2})$, $(4;\ -2\sqrt{2})$, (3; 5) and (3; −5).

23. 3; $\frac{2}{3}$ and −2 − 1.

24. 2; $\frac{3}{2}$ and $\frac{1}{2}$; $\frac{2}{3}$.

25. (7; 3), (7; −4), (−8; 3) and (−8; −4).

26. $\frac{a}{\sqrt{a+b}}$; $\frac{b}{\sqrt{a+b}}$ and $-\frac{a}{\sqrt{a+b}}$; $-\frac{b}{\sqrt{a+b}}$.

27. $(a;\ a)$, $(-a;\ -a)$, $\left[\frac{a}{2}(1+i\sqrt{3});\ \frac{a}{2}(1-i\sqrt{3})\right]$ and $\left[\frac{a}{2}(1-i\sqrt{3});\ \frac{a}{2}(1+i\sqrt{3})\right]$.

28. $a+1$; $a-1$ and $-1-a$; $1-a$.

29. $(a;\ a+b)$, $(a;\ -b)$, $(b+\sqrt{a^2+b^2};\ b)$ and $(b-\sqrt{a^2+b^2};\ b)$.

30. $(-a;\ 2a-2)$, $(-a;\ -2a)$, $(2a;\ a)$ and $(-a;\ -2a)$.

31. 7; 5 and 5; 7.

32. (3; 2), (3; −2), (−3; 2) and (−3; −2).

33. 7; 5 and 5; 7.

34. 6; 2.

35. 20; 4 and −20; −4.

36. 2; 2 and 2; 2.

37. (4; 3), (−4; −3), (3; 4) and (−3; −4).

38. 7; 6 and −6; −7.

39. 8; 2.

40. 4, 64; 64, 4.

41. (3; 2), (−3; 2), $(\sqrt{-2};\ -9)$ and $(-\sqrt{-2};\ -9)$.

42. 4; 3 and −3; −4.

43. (3; 1), (−3; −1), $(i;\ 3i)$ and $(-i;\ -3i)$.

44. 8; 4 and 4; 8.

45. (3; 1), (−3; −1), $\left(\frac{3}{2};\ 2\right)$ and $\left(-\frac{3}{2};\ -2\right)$.

46. (2; 1), (−2; −1), (i; $2i$) and ($-i$; $-2i$).

47. (3; 2), (−2; −3).

48. $\pm 3\sqrt{2};\ \pm\sqrt{2}$.

49. (9; 4), (−4; −9).

50. (9; 4), (4; 9), $(4+\sqrt{15};\ 4-\sqrt{15})$ and $(4-\sqrt{15};\ 4+\sqrt{15})$.

51. (0; 0), $\left(\frac{3+i\sqrt{3}}{2};\ \frac{3-i\sqrt{3}}{2}\right)$ and $\left(\frac{3-i\sqrt{3}}{2};\ \frac{3+i\sqrt{3}}{2}\right)$.

52. (0; 0), (1; 1) and (−1; −1).

53. (3; 2), (2; 3), $\left(\frac{5+i\sqrt{151}}{2};\ \frac{5-i\sqrt{151}}{2}\right)$ and $\left(\frac{5-i\sqrt{151}}{2};\ \frac{5+i\sqrt{151}}{2}\right)$.

54. (2; −1), $\left(\frac{3+i\sqrt{17}}{2};\ \frac{-3+i\sqrt{17}}{2}\right)$ and $\left(\frac{3-i\sqrt{17}}{2};\ \frac{-3-i\sqrt{17}}{2}\right)$.

55. (3; 1), (1; 3), $\left(2+2\sqrt{\frac{19}{7}};\ 2-2\sqrt{\frac{19}{7}}\right)$ and $\left(2-2\sqrt{\frac{19}{7}};\ 2+2\sqrt{\frac{19}{7}}\right)$.

56. (2; 1), $\left(\frac{1}{3};\ -\frac{2}{3}\right)$, $\left(\frac{1+\sqrt{41}}{10};\ \frac{-9+\sqrt{41}}{10}\right)$ and $\left(\frac{1-\sqrt{41}}{10};\ \frac{-9-\sqrt{41}}{10}\right)$.

57. (4; 2), (2; 1), (4; 2) and (2; 1).

58. (6; 3), $\left(-3,\ -\frac{3}{2}\right)$, $\left(\frac{12+3\sqrt{39}}{23};\ 12+3\sqrt{39}\right)$ and $\left(\frac{12-3\sqrt{39}}{23};\ 12-3\sqrt{39}\right)$.

59. 5; 3.

60. (333; 111) and (115; 329).

61. ±3; ±4; ±2.

62. (8; 6; 9) and (−8; −6; −9).

63. ±1; ±4; ±6.

64. (9; 3; 0) and (1; −5; −8).

65. ±5; ±2; ±7.

66. 13; 5; 2 and 2; 5; 13.

67. (4; 3; 5), (3; 4; 5), $\left(\frac{5+i\sqrt{23}}{2};\ \frac{5-i\sqrt{23}}{2};\ 7\right)$ and $\left(\frac{5-i\sqrt{23}}{2};\ \frac{5+i\sqrt{23}}{2};\ 7\right)$.

68. 2; 4; 1 and 2; 1; 4.

69. 5; 12; 13 and 12; 5; 13.

70. (1; 1; 1), (1; 1; 1), $\left(\frac{7+i\sqrt{15}}{2};\ -4;\ \frac{7-i\sqrt{15}}{2}\right)$ and $\left(\frac{7-i\sqrt{15}}{2};\ -4;\ \frac{7+i\sqrt{15}}{2}\right)$.

71. (4; 6; 3), (4; 3; 6), $(9;\ 2+i\sqrt{14};\ 2-i\sqrt{14})$ and $(9;\ 2-i\sqrt{14};\ 2+i\sqrt{14})$.

72. ±1; ±5; ±2.

73. $\frac{1}{2};\ \frac{1}{6};\ \frac{1}{4}$ and $\frac{1}{4};\ \frac{1}{3};\ \frac{1}{5}$.

74. $\frac{1}{4};\ \frac{1}{3};\ \frac{1}{2}$ and $\frac{1}{8};\ -\frac{1}{5};\ -\frac{1}{2}$.

75. (1; 2; 3), (1; 3; 2), (2; 1; 3), (2; 3; 1), (3; 1; 2) and (3; 2; 1).

76. (5; −2; −3), (5; −3; −2), (−2; 5; −3), (−2; −3; 5), (−3; 5; −2) and (−3; −2; 5).

77. 2; 5; 4; 3 and −7; 4; 5; 12.

78. (3; 5; 4; 7), (17; 5; 4; −7), $(10+\sqrt{58}; -4; -5; -\sqrt{58})$ and $(10-\sqrt{58}; -4; -5; \sqrt{58})$.

79. 10; 6; 5; 3 and 3; 5; 6; 10.

80. 3; 2; 6; 1 and 2; 3; 1; 6.

81. $\frac{1}{2}\left(a^2+2b\pm a\sqrt{a^2+4b}\right)$; $\frac{1}{2}\left(a^2+2b\mp a\sqrt{a^2+4b}\right)$.

82. $\frac{a}{\sqrt[3]{a^2+b^2}}$; $\frac{b}{\sqrt[3]{a^2+b^2}}$.

83. $\pm a$; $\pm 2a$; $\pm 3a$.

84. $\pm\frac{m}{\sqrt{3bc}}$; $\pm\frac{m}{\sqrt{3ac}}$; $\pm\frac{m}{\sqrt{3ab}}$.

85. $\pm\frac{\sqrt{a^2+b^2}}{n}$; $\pm\frac{an}{\sqrt{a^2+b^2}}$; $\pm\frac{bn}{\sqrt{a^2+b^2}}$.

86. $\pm\sqrt{\frac{(a+c-b)(a+b-c)}{2(b+c-a)}}$; $\pm\sqrt{\frac{(b+c-a)(a+b-c)}{2(a+c-b)}}$; $\pm\sqrt{\frac{(b+c-a)(a+c-b)}{2(a+b-c)}}$.

87. $\frac{a(b+c)}{2\sqrt{abc}}$; $\frac{b(a+c)}{2\sqrt{abc}}$; $\frac{c(a+b)}{2\sqrt{abc}}$.

88. $\pm\sqrt{\frac{a+b-c}{2}}$; $\pm\sqrt{\frac{a-b+c}{2}}$; $\pm\sqrt{\frac{b+c-a}{2}}$.

89. $a+b\mp\sqrt{2ab}$; $-b\pm\sqrt{2ab}$; $-a\pm\sqrt{2ab}$.

90. $m+n$; $m-n$; m.

91. 19 and 11.

92. (15; 13), (−13; −15).

93. 8 and 5.

94. 15; 5.

95. 6 and 8 or −9 and −7.

96. 24.

97. 12 and 4.

98. 13 and 9.

99. (10; 7), (−7; −10).

100. $\frac{3}{4}$ or $\frac{4}{3}$.

101. 35 or 53.

102. 20; 18; 16.

103. 1; 4; 7.

104. 452.

105. 9; 6; 4.

106. 864.

107. $\frac{1\pm\sqrt{2(a+b)+1}}{2}$ and $\frac{1\pm\sqrt{2(a-b)+1}}{2}$.

108. $\frac{a(p+q)}{2\sqrt{apq}}$; $\frac{a(p-q)}{2\sqrt{apq}}$.

109. 14 and 8.

110. 24; 4.

111. 36 and 12.

112. 5; 12; 13.

113. 6; 8; 10.

114. 5; 12; 13.

115. 5; 6; 8.

116. 9; 3.

117. $\frac{\sqrt{a^2+4s}\pm\sqrt{a^2-4s}}{2}$.

118. $a = \frac{4ps}{p^2+s\pm\sqrt{(p^2+s)^2-8p^2s}}$; $b = \frac{p^2+s\pm\sqrt{(p^2+s)^2-8p^2s}}{2p}$; $c = 2p - a - b$.

119. $\frac{4p\pm d-\sqrt{8p^2+d^2}}{2}$ and $\sqrt{8p^2+a^2} - 2p$.

120. $\sqrt{a^2+s} \pm \sqrt{a^2-s}$.

121. 120 sec.; 0.2 m/sec.^{2}.

122. 12 and 9.

123. 45 and 25.

124. 16 and 12.

125. 12 and 5.

126. 24; 33.2.

129. $\sqrt{pq}$.

131. $\frac{(ad-cb)ap}{(a+b)dq-(c+d)bp}$; $\frac{(bc-ad)pq}{(a+b)cq-(c+d)ap}$.

132. $\frac{a}{\sqrt{2a-d}}$; $\frac{d-a}{\sqrt{2a-d}}$.

133. $\frac{d+\sqrt{a^2+4md}}{2md}$; $\frac{-d+\sqrt{a^2+4md}}{2md}$.

134. $\frac{a(m^2+c^2)-2bcv\pm(v-c)\sqrt{a^2(v+c)^2-4abcv}}{2c}$; $\frac{a(v-c)\pm\sqrt{a^2(v+c)^2-4abcv}}{2a}$.

Chapter XV. Progressions

1. 44; 345.

2. −37; −360.

3. 1065.

4. $\frac{n(n+1)}{2}$.

5. 5050.

6. 10100.

7. $2n-1$; n^2.

8. $\frac{a(n+1)-b(n-1)}{2}n$.

9. $d = 3$.

10. $d = -5$.

11. 55; 403.

12. 26; 451.

13. 2; 1661.

14. 56; 680.

15. 5; 18.

16. −1; 20.

17. 4; 528.

18. −2; 330.

19. 140; 10.

20. 45; 3.

21. 9; 2.

22. 0; 7.

23. 10; 265.

24. 26; 604.5.

25. 7; 61.

26. −9; 3.

27. 10; 47.

28. 52; 143.

29. 10; 2.

30. 24 and −4 or 21 and 8.

31. 33; −4.

32. 145.

33. 9 or 4.

34. 10; 8; 6...

35. −10.

36. 1, 3, 5... 21.

37. 1, 3, 5...

38. $\frac{(m+n)(m+n-1)}{2}$.

39. 2, 5... or 5, 2...

40. −100.

41. 2, 5, 8... or 8, 5, 2...

42. 2, 5, 8... or 14, 11, 8...

43. 8 or 9.

44. 9 or 16.

45. 200.9; 2160.9.

46. 30.

47. 67.

48. 160.

49. 441 m.

50. 4410 m; 294 m.

51. 6 cm; 48 cm.

52. 1700 m.
53. 3.
54. 208, 176, 144, 112.
55. 7.
56. 10 m.
57. 21 and 80 sec.
58. 15 sec.
59. 2745 m; 99 km; 36.3 km; 206 000°.
60. 5 months.
61. 78.4 m.
62. Approximately 4.5 sec.
63. 4.
64. 4.
65. 40 m.
66. 10 230.
67. −13 108.
68. $\frac{1640}{729}$.
69. $\frac{19\sqrt{6}+30}{12}$.
70. $\frac{8}{3}\left[1-\left(\frac{3}{4}\right)^n\right]$.
71. $\frac{\sqrt{6}\left(\sqrt{3^n}-1\right)}{\sqrt{3}-1}$.
72. 141 and 423.
73. $q=\sqrt{\frac{b}{a}}$.
74. 189.
75. $\frac{b^{m+1}}{a+b}\cdot\left[\left(-\frac{a}{b}\right)^n-1\right]$.
76. 13 122; 19 680.
77. $-\frac{1}{625}$; $4\frac{104}{625}$.
78. 2; 254.
79. $\frac{3}{8}$; $\frac{55}{216}$.
80. 8; 14 043.
81. $-\frac{2}{3}$; $\frac{133}{3}$.
82. 5; 320.
83. 8; $-\frac{1}{16}$.
84. 6; 189.
85. 6; $\frac{665}{27}$.
86. 3; 7.
87. 2; 6.

88. 567; 5.

89. -486; 6.

90. -6 and 3.

91. 2; 8.

92. 2; 120.

93. -6; 432 or 5; 300.

94. $-\frac{2}{3}$; 27.

95. 3; 15 or $-\frac{3}{4}$; 240.

96. $q = \pm 3$; $\pm\sqrt{-10}$.

97. 3069.

98. 54; 18; 6 and 2 or 27; -9; 3 and -1.

99. 2; 14; 98.

100. 64; 32; 16; 8; 4 and 2.

101. 2; 6; 18.

102. 5; 13; 21; 29.

103. $\sqrt{kl}$ and $k\sqrt[2n]{\left(\frac{l}{k}\right)^m}$.

105. $\frac{\sqrt{2}}{256}$.

106. 15.5; 22.0...247.6.

107. Minimum 9.07.

108. 9 m; 12 m; 16 m.

109. 18 m; 6 m; 2 m.

110. Approximately 46.5 mm.

111. 26.7 L.

112. 625 L.

113. 2.

114. $\frac{3}{4}$.

115. $\frac{3\sqrt{6}}{2}$.

116. $\frac{16+11\sqrt{2}}{7}$.

117. $q = \frac{1}{1+k}$.

118. $\frac{1}{3}AB$.

119. $4a\left(2+\sqrt{2}\right)$ and $2a^2$.

120. $6x\left(2-\sqrt{3}\right)$ and $\frac{1}{7}a^2\sqrt{3}$.

121. $2\pi R^2$ and $4R^2$.

122. $\left(1+\sqrt{2}\right)a$.

Chapter XVI. Logarithms

1. 8.
2. 2.
3. 2.
4. 64.
5. 4.
6. $\frac{1}{4}$.
7. 4.
8. 64.
9. $\frac{1}{25}$.
10. -3.
11. 10; 5; 2.
12. -4; -2; -1.
13. $\frac{1}{512}$.
14. 3.
15. -6; -3; -2.
16. 6; 3; 2.
17. $1; \frac{3}{4}; \frac{4}{3}; \frac{9}{16}; \frac{16}{9}; \frac{27}{64}; \frac{64}{27}$.
18. -1; 2; 0; -3.
19. $5; 125; \frac{1}{5}; \frac{1}{125}$.
20. 0; -2; -1; 2; -3; 3.
21. $\sqrt[4]{27}$.
22. 1 and 2; 2 and 3; 3 and 4; 4 and 5; 6 and 7; 8 and 9.
23. -1 and -2; -1 and -2; -2 and -3; -3 and -4; -4 and -5.
24. $\sqrt{5}$.
25. $\frac{\sqrt[4]{8}}{8}$.
26. $\frac{\sqrt[3]{7}}{7}$.
27. $-\frac{1}{8}; -512; \frac{1}{64}; -2; -\frac{1}{2}$.
28. 3; 2; -4.
29. $\frac{2}{5}; \frac{1}{10}; -\frac{1}{5}; -\frac{1}{10}$.
30. $4; \frac{1}{\sqrt{2}}; \frac{\sqrt{2}}{4}; 4\sqrt[4]{2}$.
31. 0.
32. ∞.
33. $-\infty$.
34. $\log_{14} 15 > \log_{15} 14$.
35. $\log 2 + \log 3$.
36. $\log 5 - \log 3$.

37. $3\log 5$.

38. $\frac{1}{4}\log 11$.

39. 9; -1.

40. 2 and 3; 3 and 5; 2 and 19; 5 and 7.

41. 0.77815; 1.17609; 1.47712; 1; 3.

42. 0.39794; 0.22185; −1.09691; −0.22185; −1.79588.

43. 1.30103; 2.30103; 1.17609; 2.17609; 3.17609.

44. −0.52288; −2.52288; −1.22185; −3.22185.

45. $\log 2 + \log a + \log b$.

46. $\log a + \log b - \log c$.

47. $3\log a + 2\log b$.

48. $2\log a - 3\log b - 7\log c$.

49. $\log 2 + \log(a+b)$.

50. $\log 3 - \log(a+b) - \log(a-b)$.

51. $2\log(a-b) + \log c - \log(a+b) - \log d$.

52. $\log 5 + 2\log a + \log b + \frac{1}{3}\log c$.

53. $\frac{1}{2}(\log 3 + 3\log a + \log b - 4\log c)$.

54. $\log 5 + \frac{1}{3}[5\log a + \log(a-b)]$.

55. $\log 2 + \log a + 3\log b - \log c - \frac{1}{2}\log d$.

56. $\frac{5}{3}(1 - \log a)$.

57. $-n\log a - \frac{1}{2}\log b$.

58. $-\log a - \frac{1}{2}\log b$.

59. $\frac{2}{3}\log a + \frac{3}{5}\log b$.

60. $\frac{m}{n}(\log m - \log n)$.

61. $\frac{1}{2}\left[\log 2 + \frac{1}{2}\left(\log 6 + \frac{1}{2}\log 15\right)\right]$.

62. $\frac{1}{3}\left(2\log a + \log b - \frac{3}{5}\log c\right)$.

63. $-\frac{3}{4}\log a + 2\log b + \frac{1}{5}\log c$.

64. $\frac{1}{2}\left[\log 24 + \frac{1}{2}\left(\log 2 + \frac{1}{2}\log 3\right) - \frac{1}{3}\left(\log 4 + \frac{1}{2}\log 6\right)\right]$.

65. 0.

66. $\frac{1}{m}\left(\log a + \frac{1}{n}\log b\right)$.

67. $-\frac{1}{m}\left(\log a - \frac{1}{n}\log b\right)$.

68. $\frac{\sqrt{2}}{2}\log 2$.

69. $-\frac{1}{2}\log 2 + \log\log 2$.

70. $\log a + \log b + \log\log 10$.

71. $\log\frac{10}{3}$.

72. $2\log\log(a+b)$.

73. $\log 4 - \log 5 + \frac{2}{3} \log a + \log \log a.$

74. $\log \log(a+b) + \log \log(a-b) - \log 2.$

75. $\frac{14}{3}.$

76. 1125.

77. $\frac{\sqrt[5]{11^3}}{\sqrt[7]{25}}.$

78. $\frac{169}{\sqrt[5]{4} \cdot \sqrt[3]{7^4}}.$

79. $\frac{a^3 b^2}{c^4}.$

80. $\frac{\sqrt{a}\,\sqrt[3]{b}}{\sqrt[4]{c}}.$

81. $\frac{\sqrt{10}}{\sqrt[3]{5}}.$

82. $\frac{\sqrt[5]{(a+b)^2}}{\sqrt[4]{(a-b)^3}}.$

83. $\frac{a+b}{\sqrt[3]{(a^2 \sqrt[4]{b^3})^2}}.$

84. $\frac{\sqrt[3]{\frac{(a+b)\sqrt[5]{(a-b)^2}}{b\sqrt{c}}}}{a^3}.$

85a. 1.30103; 3.30103; $\overline{1},30103$; $\overline{6},30103$.

86. 0.39794; 2.69897; $\overline{1},39794$; $\overline{3},69897$.

87a. 2.32222; $\overline{2},32222$.

88. 0.17609; $\overline{1},07918$; $\overline{1},55630$.

89. 0.90309; 2.14922; 2.97955; 2.62325; 2.80618; 3.09167; 3.59184; 3.47857; 1.26553; 0.31175; 2.95429; $\overline{1},86332$; $\overline{3},44716$; $\overline{1},00346$; $\overline{5},69897$.

91. 1453; 3909; 857.4; 2.575; 39.87; 0.05713; 0.001719; 0.6265; 0.0007552; 0.000014.

92. 3774.75; 1456.8; 253.272; 12.5365; 31494.3; 0.0068995; 0.316664; 0.0019978; 0.148293; 0.00017824.

93. $\overline{3},30463$; $\overline{5},78707$; $\overline{1},45775$; $\overline{2},31693$; $\overline{4},46180$; $\overline{6},10010$.

94. −0.66722; −2.47207; −1.04574; −3.76275; −0.60580; −4.32010.

95. 7961.6.

96. 401.74.

97. 31.

98. 41.846.

99. 552.25.

100. 0.000021952.

101. 3.5355.

102. 0.37325.

103. 40.42.

104. 0.18894.

105. 1.4252.

106. 0.7372.
107. 5.55562.
108. 0.13762.
109. 50.466.
110. 1.04712.
111. 0.37077.
112. 0.00068129.
113. 4.8674.
114. 1.0295.
115. 0.050187.
116. 0.79668.
117. 93.832.
118. 0.1567.
119. 0.77738.
120. 0.85856.
121. 0.018886.
122. 0.146143.
123. 1.24203.
124. 0.90084.
125. 0.8233.
126. 3.18957.
127. 79 568.
128. 21.55.
129. −8094.66.
130. 2.8946.
131. 248.4.
132. 0.0025534.
133. 0.13762.
134. 10.4864.
135. 0.9937.
136. 0.88396.
137. 0.88662.
138. 0.537275.
139. 13.715 cm; 41.145 cm.
140. 4.4247; 7.3745; 11.7992.
141. 8.6556 cm.
142. 2.9961.
143. 696.55 cm^3.

144. 3.275.
145. 0.7937 m.
146. 11 cm.
147. 16.453 cm.
148. 2.8405.
149. $9°54'$.
150. 64.4 kg.
151. $6 \cdot 10^{24}$.
152. 3.188.
153. 592.3 m.
154. 30.62.
155. 2.04 dm.
156. 7.871.
157. 512 m.
158. 21.99 L.
159. 99.45 cm.
160. 74.87.
161. 1.3631.
162. 2.96807.
163. 0.4618.
164. 73.23.
165. 1.199.
166. -25.3944.
167. 1.15740.
168. 3.42838.
169. 1.596.
170. -0.88853.
171. 0.093428.
172. 0.85119.
173. 1.16327.
174. 1.80517.
175. -4.
176. $-\frac{6}{7}$.
177. $-\frac{1}{2}$.
178. 7.
179. $\frac{5}{2}$.
180. $-\frac{2}{5}$.
181. $\frac{7}{3}$.

182. 4 and -1.
183. ± 2.
184. 2.
185. 2.
186. 3.
187. 2 and 5.
188. $\frac{1}{a+b}$.
189. 35.
190. 3; 1 and -2.
191. 5.2 and 2.5.
192. 1.5 and -0.5.
193. 1000.
194. 3 and -5.
195. 3.
196. 4.
197. 0.
198. 1.
199. 2.
200. 1.
201. 1.
202. 0; 1 and 4.
203. 2 and -1.585.
204. 10 and 0.1.
205. 100 and 0.1.
206. 1000 and 0.01.
207. $\pm\sqrt{\lg 5} \approx \pm 0.836$.
208. 1.76.
209. 2.3.
210. -5.13.
211. 1.47.
212. 0.54866.
213. -0.58275.
215. -0.43683.
216. 1.1893.
217. -7.73977.
218. 18.
219. $3\frac{1}{3}$.
222. 25.

223. $\frac{16}{\sqrt[3]{5}}$.

224. $\frac{1}{2}$; $4\frac{1}{2}$.

226. 4.

227. 9; 1.

228. ± 2.

229. $\frac{-1\pm\sqrt{5}}{2}$.

230. 13; $-3\frac{1}{2}$.

231. 18.

232. a^{mn}.

233. $\sqrt[n]{m}$.

234. $\log_a(b \pm \sqrt{b^2 - c^2})$.

235. 2.

236. -2.

237. 1000; 0.001; 1; 0.1.

238. $\frac{2}{3}$ and 1.969.

239. 16 and 10.

240. 1 000 000 and 10.

241. 1.6624 and 1.2745.

242. $\frac{27}{8}$ and $\frac{9}{4}$.

243. 4 and 2 or 9 and -3.

244. 2.25 and 0.75.

245. 2 and 4.

246. 1 and 1 or 16 and 4.

247. 3 and 5.

248. 3 and 2.

249. 363 rubles 47 kopecks.

250. 2493 rubles 94 kopecks.

251. 20.

252. 4%.

253. 7.18%.

254. 5.5%.

255. Approximately 20 years.

256. 4543 rubles 12 kopecks.

257. Approximately 23 years.

258. 19 675.26 rubles.

259. 11 603.3 rubles.

260. With an insignificant increase in kopecks.

261. 6 years.

262. Approximately by 750 m^3.

263. Approximately 3.3%.

264. Approximately 31 900 rubles.

265. Approximately $28 \cdot 10^{14}$.

266. 8.85%.

267. 79 948 m^3.

268. 20 454 m^3.

269. 17 years.

Chapter XVII. Combinatorics

1. *abc*; *acb*; *bac*; *bca*; *cab*; *cba*.

2. *abc*; *acb*; *bac*; *bca*; *cab*; *cba*; *abd*; *acd*; *bad*; *bcd*; *cad*; *cbd*; *adc*; *adb*; *bda*; *cab*; *cda*; *dbc*; *dcb*; *dac*; *dca*; *dab*; *dba*.

3. *abcd*; *abc*; *abd*; *bca*; *acd*; *ab*; *ac*; *ad*; *bc*; *bd*; *cd*; *a*; *b*; *c*; *d*.

7. 210; 120; 15.

8. 40 320; 8 648 640; 54 264.

9. $(n+1)n(n-1)\dots(n-k+3)$.

10. $(m+n)(m+n-1)(m+n-2)\dots 2n$.

13. $\frac{(n+2)(n+1)n\dots(n-k+4)}{1\cdot 2\cdot 3\dots(k-1)}$.

14. $\frac{(m-n)(m-n-1)(m-n-2)\dots(m-2n)}{1\cdot 2\cdot 3\dots(n+1)}$.

15. 24.

16. 840.

17. 3024.

18. 45.

19. 15.

20. 6.

21. 14 or 3.

22. 7.

24. 24; 6; 2.

25. 84; 28.

26. 7920; 720.

27. C_{h-n}^{k-n}.

28. A_{n-h}^{k-h}.

29. $k < \frac{n+1}{2}$; $\frac{n-1}{2}$ or $\frac{n}{2}$.

Chapter XVIII. Newton's Binomial

1. $x^4 + 10x^3 + 35x^2 + 50x + 24$.

2. $x^4 + 3x^3 - 21x^2 - 43x + 60$.

3. $x^5 + 15x^4 + 85x^3 + 225x^2 + 274x + 120$.

4. $x^5 - 4x^4 - 37x^3 + 124x^2 + 276x - 720$.

5. $a^6 + 6a^5b + 15a^4b^2 + 20a^3b^3 + 15a^2b^4 + 6ab^5 + b^6$.

6. $a^7 - 7a^6b + 21a^5b^2 - 35a^4b^3 + 35a^3b^4 - 21a^2b^5 + 7ab^6 - b^7$.

7. $a^9 + 9a^8 + 36a^7 + 84a^6 + 126a^5 + 126a^4 + 84a^3 + 36a^2 + 9a + 1$.

8. $1 - 8a + 28a^2 - 56a^3 + 70a^4 - 56a^5 + 28a^6 - 8a^7 + a^8$.

11. $a^3 + 6a^2\sqrt{ab} + 15a^2b + 20ab\sqrt{ab} + 15ab^2 + 6b\sqrt{ab} + b^3$.

13. $126a^5b^4$.

14. $-3432a^7b^7$.

15. $C_{19}^{8}a^8x^{11}$ and $C_{19}^{11}a^{11}x^8$.

16. $C_{24}^{6}a^6x^{42}$ and $C_{24}^{6}a^{18}x^{30}$.

17. $84z^4$.

18. $\frac{1120}{a^4}$.

19. $715(1-z)^2(1+z)^4\sqrt{1+z}$.

Chapter XIX. Divisibility of Polynomials

1. Without remainder.

2. With a remainder.

3. Without remainder.

4. Without remainder.

5. Without remainder.

6. With a remainder.

7. With a remainder.

8. 0.

9. 23.

10. 0; $\frac{1 \pm i\sqrt{23}}{2}$.

11. 1; $1 \pm i\sqrt{3}$.

12. 1; 2; -3.

13. -3; $\frac{1 \pm i\sqrt{3}}{2}$.

14. 1; -2; $\pm i\sqrt{2}$.

15. $(x+1)(x-2)(x-3)$.

16. $(x+1)(x+2)(x-2)(x-4)$.

17. $(x+1)(x-1)^2(x-2)$.

Chapter XX. Inequalities

1. $13 > 2.$

3. $x^2 + 2x > 2a - 4.$

5. $14 > 8.$

6. $-6 < 2.$

7. $x - a^2 > b^2 - 9.$

8. $-4ab < -6.$

9. $25 > -10.$

10. $14 > 10.$

11. $-a^2b < -b^2.$

13. $-2 < 3.$

14. $3 < 7.$

15. $-a^2 > -a.$

16. $(a-b)^3 > a - b.$

17. $35 > 6.$

18. $-6 < 35.$

19. $15 > 10.$

21. $5 < 8.$

22. $-2 < 2.$

24. $-4\frac{4}{35} < -3.$

25. $x > -\frac{1}{2}.$

26. $x < -2.$

27. $x > \frac{24}{25}.$

28. $x > 56.$

29. $x < -\frac{4}{5}.$

30. $x < -3\frac{1}{2}.$

31. $x > 8.$

32. $x < 1\frac{2}{3}.$

33. $x > 10\frac{2}{3}.$

34. $x < 2.$

35. $x > -3\frac{2}{3}.$

36. $x < -5.$

37. $x > \frac{1}{2}.$

38. $x > 7\frac{1}{2}.$

39. $x < \frac{4}{5}.$

40. $x < \frac{1}{5}.$

41. $x < -3.$

42. $1 < x < 4.$

43. $x > \frac{3}{2}$.
44. $3 < x < 19$.
45. Incompatible.
46. Incompatible.
47. $x > -2$.
48. $a < \frac{2}{3}$ or $a > \frac{3}{2}$.
49. $-3\frac{1}{2} < a < \frac{2}{3}$.
50. $a < \frac{2}{7}$ or $a > 2\frac{2}{3}$.
51. $-1\frac{3}{5} < a < 2\frac{1}{3}$.
59. $n < \frac{q}{p}(a + m) - b$.
60. $\frac{m}{n} > \frac{p}{N} + 1$.
61. $43.6 \geqslant x \geqslant 34.6$.
62. $29.4 > x > 22.2$; $27.8 > y > 20.6$.
63. $V \leqslant 366\frac{2}{3}$.

Chapter XXI. Indeterminate Equations of the First Degree

1. $7 - 2t$; t.
2. t; $12 + 5t$.
3. $5t$; $3t$.
4. $8t$; $-5t$.
5. $5 + 3t$; $1 - 2t$.
6. $5t - 3$; $7t$.
7. $12 - 13t$; $7t - 1$.
8. $4 + 9t$; $5 + 14t$.
9. $2 + 3t$; $1 - 2t$.
10. $1 + 4t$; $3t - 2$.
11. $3t$; $2 - 5t$.
12. $4t - 3$; $7t - 6$.
13. $1 - 5t$; $1 + 7t$.
14. $11t - 8$; $5t - 4$.
15. $3 - 8t$; $5 + 11t$.
16. $1 + 7t$; $6 + 11t$.
17. No.
18. No.
19. No.
20. No.

21. No.

22. Yes.

23. No.

24. Yes.

25. 9 and 1.

26. 9; 16 . . . and 9; 21 . . . .

27. 6 and 3.

28. 4; 53 . . . and 1; 16 . . . .

29. 25; 60 . . . and 12; 30 . . . .

30. 2 and 1.

31. 5; 15; 25; 35; 45; 55 and 51; 42; 33; 24; 15; 6.

32. 0; 5; 10; 15; 20 and 28; 21; 14; 7; 0.

33. 7; 11 . . . and 9; 24 . . . .

34. 11; 3.

35. 14; 12.

36. 5; 2.

37. 11; 7.

38. 23; 21.

39. 23; 17.

40. (15; 4; 3), (30; 11; 7).

41. (2; 2; 1).

42. (0; 7; 7), (1; 6; 5).

43. (7; 2; 0), (6; 3; 1).

44. (18; 3; 1), (73; 14; 6).

45. (2; 3; 4).

46. (34; 22; 5).

47. 70 and 130 or 161 and 39.

48. 10 ways.

49. $136t - 24$ and $136t - 34$.

50. 7 solutions or infinitely many.

52. 15 and 10 or 6 and 26.

53. 22 and 25 or 4 and 50.

54. The numerators of the first 5.8 . . ., second 2.6.

55. $5 + 24t$.

56. $40t + 25$.

58. $-21 - 40t$.

59. $17 + 21t$.

60. $11 + 56t$.

61. 29 and 5, or 56 and 13, or 83 and 21.

62. In the first case, the numbers of revolutions are in the ratio 23:19, in the second solution 6.29... and 5.24..., in the third solution 12.35... and 10.29....

65. 58.

66. $1320t + 25$.

68. 4, 4, 1; 1, 6, 1; 3, 3, 2; 6, 1, 2; 2, 2, 3; 1, 1, 4.

69. 22; 8, 45.

Chapter XXII. Continued Fractions

1. $\frac{62}{23}$.

2. $\frac{201}{88}$.

3. $\frac{37}{104}$.

4. $\frac{a^3b^2+4a^2b+3a}{a^2b^2+3ab+1}$.

5. $\frac{6x^3+5x}{6x^4+7x^2+1}$.

6. $\frac{a^4+2a^2-a+1}{a^3+a^2+2a}$.

7. (2, 7, 1, 6).

8. (0, 1, 5, 3, 7).

9. (2, 2, 2, 2, 2, 3).

10. (0, 2, 1, 2, 1, 1, 4, 2).

11. $(a, a-1, a+1, a)$.

12. $0, \frac{1}{2}, \frac{2}{5}, \frac{5}{12}, \frac{12}{29}, \frac{29}{70}, \frac{99}{239}$.

13. $0, 1, \frac{1}{2}, \frac{2}{3}, \frac{3}{5}, \frac{5}{8}, \frac{8}{13}, \frac{13}{21}, \frac{21}{34}, \frac{34}{55}, \frac{55}{89}$.

14. $0, \frac{1}{2}, \frac{3}{7}, \frac{10}{23}, \frac{33}{76}, \frac{43}{99}, \frac{76}{175}, \frac{271}{624}, \frac{1702}{3919}$.

15. $1, \frac{4}{3}, \frac{21}{16}, \frac{151}{115}, \frac{1380}{1061}, \frac{15331}{11676}$.

16. $0, \frac{1}{10}, \frac{100}{1001}, \frac{100001}{1001010}$.

17. (1, 2, 2...).

18. (1, 1, 2, 1, 2...).

19. (4, 2, 8, 2, 8...).

20. (2, 1, 1, 1, 4, 1, 1, 1, 4...).

21. $(a, 2a, 2a \ldots)$.

22. $\sqrt{17}$.

23. $\sqrt{15}$.

24. $\frac{\sqrt{15}-3}{2}$.

25. $\sqrt{a^2+a}$.

26. $5 - 13t$; $8t - 3$.

27. $14t - 9$; $9t - 6$.

28. $14 - 16t$; $23t - 20$.

29. $11t + 8$; $7t + 5$.

30. $34t - 20$; $29 - 49t$.

31. $(2, 1, 2, 3 \ldots) \approx 2.7$.

32. $(3, 1, 1, 1 \ldots) \approx \frac{11}{3}$.

33. $(1, 2, 2, 1, 1, 2, 2 \ldots)$.

34. $(1, 1, 1, 2, 3, 9 \ldots)$.

35. $(2, 10, 1, 1 \ldots)$.

36. $(0, 1, 1, 3 \ldots)$.

Chapter XXIII. Investigation of Equations

1. $a > 2\frac{1}{2}$.

2. $a < 3$.

3. $0 < a < 5$.

4. $5 < a < 8$.

5. $9 > a > 7$.

6. $a < 2\frac{2}{3}$ or $a > 7\frac{1}{2}$.

15. Impossible.

16. Impossible.

17. $\frac{a-b}{n-m}$.

18. $\frac{a-bk}{1-k}$.

19. $\frac{a-bk}{k-1}$.

20. $\frac{ab}{b-a}$.

21. $\frac{an-bm}{m-n}$.

22. $\frac{b-am}{m}$.

23. $\frac{a(q-n)}{m-n+q-p}$.

24. $\frac{bc}{b-a}$.

25. $a > 3\frac{1}{3}$.

26. $-4 < a < 3\frac{3}{4}$.

27. $-107\frac{1}{3}$.

28. $a = -17.5$.

31. $0 < b < \frac{a^2}{4}$.

32. $b^2 < a^2 < 2b^2$.

33. $\frac{1}{2}(\sqrt{a^2 + 4b^2} \pm \sqrt{a^2 - 4b^2})$; $\frac{1}{2}(\sqrt{a^2 + b^2} \mp \sqrt{a^2 - 4b^2})$.

35. $\sqrt{R^2 + k^2} \pm \sqrt{R^2 - k^2}$; $\sqrt{R^2 + k^2} \mp \sqrt{R^2 - k^2}$.

36. $x < 0$.

37. $-1 < x < 3$.

Chapter XXIV. Limits

1. 5.
2. $-\frac{1}{3}$.
3. $\frac{1}{3}$.
4. 0.
5. $\frac{3}{2}$.
6. 4.
7. -1.
8. $\frac{4}{3}$.
9. $\frac{3a}{2}$.
10. 3.
11. $\frac{5}{4}$.
12. 0.
13. $\frac{a}{c}$.
14. $\frac{a}{c}$.
15. 4.
16. $\frac{3}{2}$.
17. $\frac{4}{3}$.
18. $2(2+\sqrt{2})$.
19. $\frac{1}{3}$.
20. $2\frac{31}{90}$.
21. All.
22. $\frac{h^2}{h}$ and h^3.
24. $\sin h$ and $\tan h$.
25. 0; b; $\frac{b}{d}$; $\frac{b}{d}$; $\frac{b}{d}$.
26. 0; -2; $-\infty$; $\lg 5$; 1.

www.ingramcontent.com/pod-product-compliance
Lightning Source LLC
LaVergne TN
LVHW060547061126
851857LV00017B/1929
9781919012506